TECHNOLOGICAL CHANGE
AND THE TRANSFORMATION
OF AMERICA

Technological

Change and

Southern Illinois University
Carbondale and Edwardsville

THE TRANSFORMATION

OF AMERICA

Edited by

Steven E. Goldberg

and

Charles R. Strain

To our Parents
Arlene Goldberg
Richard Goldberg
Ruth Farrell Strain
Charles R. Strain

90 89 88 87   4 3 2 1

Library of Congress Cataloging-in-Publication Data

Technological change and the transformation
of America.

Bibliography: p.
Includes index.
1. Technology and civilization.   2. Technology—
Social aspects.   I. Goldberg, Steven E.   II. Strain,
Charles R.
HM221.T3977   1987   303.4'83   86-26092
ISBN 0-8093-1351-0

Contents

CONTENTS

Part Three
Community

Part Four
Politics

PREFACE

Ann Douglas has shrewdly observed that sentimentalism exists when "the values a society's activity denies are precisely the ones it cherishes." If this is true, widespread interest in "technology and values" should arouse suspicion that the many books and symposia are exercises in sentimentality, "calculated not to interfere with [our] actions." At least in the case of this volume, we believe such doubts are unfounded.

Many of the papers in this volume were originally presented at a conference, The Human Side of High Tech, sponsored by DePaul University's Institute for Business Ethics and the Department of Educational Relations, Illinois Bell in November 1984. That conference brought together representatives of education, business, labor, and government. The very composition of the conference forced all participants to encounter perspectives far different from their own. Although the papers have been extensively revised, they retain the lively tension between theory and practice that animated two days of debate.

We supplemented papers selected from the conference program with specially commissioned articles. Like the earlier conference presenters, these authors were chosen because they represent different traditions, interests, and points of view.

As our introduction makes clear, we are convinced that reasoned discourse about the ends of society can make a material difference in public policy discussions. In some distinctive way each of our authors shares this conviction. Whether this collection achieves its purpose we leave to our readers to decide.

We are happy to acknowledge the assistance of those who have helped bring this project to completion. Richard J. Meister, Dean of the College of Liberal Arts and Sciences and Brother Leo V. Ryan, C.S.V., Dean of the College of Commerce, DePaul University, have been both the inspiration behind and the stalwart supporters of the Institute for Business Ethics. Dr. Robert Allan Cooke, Director of the Institute, has consistently promoted the dialogue between representatives of education, business, government and labor which lies at the heart of this project. Mr. Patrick Keleher, Jr., President of Educational Network, Chicago, Illinois, helped shape the conference out of which this volume grew.

We appreciate the support of the Illinois Humanities Council and the National Endowment for the Humanities for both the conference and this book. Mr. Robert Klaus, Executive Director of the IHC, provided sound counsel throughout the project. We are particularly indebted to the Ameritech Foundation and its Executive Director, Mr. Michael Kuhlin, for supporting postconference events including the publication of this volume. The Committee on Faculty Research and Development of DePaul's College of Liberal Arts and Sciences awarded a grant to Charles Strain to aid in the completion of this book.

We also are indebted to the advisory council which guided the planning and execution of the initial conference and subsequent activities. The members of the council included Sarina Bellmann, Michael G. Gorman, Irving L. Kamradt, and Betsy Ann Plank of Illinois Bell; Elliott Goldstein and Michael S. Miller of DePaul University; W. Robert Cahill, Morton Thiokol; James W. Carey, University of Illinois at Urbana; John J. Hazard, Ameritech; Henry B. Hufnagel, Boeing Computer Services; R.C. Longworth, Chicago *Tribune*; F. Byron Nahser, Frank C. Nahser Advertising; Norman Peterson, Governor's Commission on Science and Technology; James P. Pitts, Northwestern University; and Charles Sengstock, Motorola.

Finally, we are most grateful to Ms. Margaret Kelton for the patience, skill, and efficiency with which she completed the many secretarial tasks connected with this project.

CONTRIBUTORS

Irving Bluestone is University Professor of Labor Studies at Wayne State University. He is a former Vice President of the International Union, UAW, and currently co-chair of the Economic Alliance for Michigan.

Albert Borgmann is Professor of Philosophy at the University of Montana. He is the author of *Technology and the Character of Contemporary Life: A Philosophical Inquiry* and of numerous articles on technology and philosophy.

Paul F. Camenisch is Professor and Chair of Religious Studies at DePaul University. He is the author of *Grounding Professional Ethics in a Pluralistic Society* and of numerous articles in business and professional ethics. He is on the board of directors for the Center for Ethics and Corporate Policy, Chicago, Illinois.

James W. Carey is the Dean of Communications at the University of Illinois, Urbana-Champaign and a former fellow of the National Endowment for the Humanities in science, technology, and human values.

Robert Allan Cooke is Associate Professor of Philosophy and Director of the Institute for Business Ethics at DePaul University. He has published articles in *The Journal of Business Ethics, Listening,* and *Case Studies in Business Ethics* (ed. Thomas Donaldsen).

Steven E. Goldberg is Special Assistant to the Dean of Faculties and Lecturer in Philosophy at DePaul University. He is the author of *Two Patterns of Rationality in Freud's Writings* (in press).

Martin G. Kalin is Associate Professor of Computer Science at DePaul University. He is the author of *Marx and Freud: The Utopian Flight from Unhappiness* and of numerous articles in the *Southern Journal of Philosophy, Kant Studien*, and other journals in both philosophy and computer science.

John F. Kasson is Professor of History and Adjunct Professor of American Studies at the University of North Carolina at Chapel Hill. He is the author of *Civilizing the Machine: Technology and Republican Values in America, 1776–1900* and *Amusing the Million: Coney Island at the Turn of the Century.*

Christopher Lasch is Watson Professor of History at the University of Rochester. He is the author of seven books including *The Culture of Narcissism,* and most recently, *The Minimal Self: Psychic Survival in Troubled Times.*

Leo Marx is the William R. Kenan Professor of American Cultural History at the Massachusetts Institute of Technology. He is the author of *The Machine in the Garden: Technology and the Pastoral Ideal in America* and has written widely in periodicals ranging from *Commentary* to *Sewanee Review.*

The Honorable Charles McC. Mathias, Jr., is the former Republican Senator for the State of Maryland.

John T. Metzger is Research Assistant at the Center for Neighborhood Development, College of Urban Affairs, Cleveland State University. During 1985–86, he served as an economic consultant to the Oakland Planning and Development Corporation. He has worked with Marc A. Weiss on several studies of urban policy and technological change.

Robert Rotenberg is Associate Professor of Anthropology at DePaul Unviersity. He has published articles in *East European Quarterly, Anthropological Quarterly, The Journal of Ethnohistory,* and *Urban Anthropology* among other journals.

Charles R. Strain is Associate Professor of Religious Studies and Director of the Master of Arts in Liberal Studies Program at DePaul University. He is the author (with Dennis McCann) of *Polity and Praxis: A Program for an American Practical Theology.*

Brian G. Sullivan is Associate Professor of Management at Western Kentucky University. Recent publications include: "Laborem Exercens: A Theological and Philosophical Foundation for Business Ethics," *Listening: Journal of Religion and Culture* (Spring 1985) and "Bang the Drum Slowly: The Demise of Collective Bargaining." *Midwest Business Law Journal* (in press).

Marc A. Weiss is Assistant Professor of Urban Planning and Policy, University of Illinois at Chicago. During 1985–86, he served as an economic consultant to the Oakland Planning and Development Corporation. He has worked with John T. Metzger on several studies of urban policy and technological change.

Technological Change
and the Transformation
of America

Introduction

Modern Technology and the Humanities

CHARLES R. STRAIN
AND
STEVEN E. GOLDBERG

Technology and the Figure of Prometheus

As told by Aeschylus, Prometheus stole from Athena and Hephaestus the gift of skill in the arts, together with fire, and shared both with his fellow mortals. "So did I lead them on to knowledge of the dark and riddling art" (Aeschylus 520). For his hubris, his arrogation of divine powers, Prometheus was punished. Aeschylus's version of Prometheus leaves us with two lessons worth considering. First, the very inventiveness which empowers man to know and master nature also can undermine him. Second, because man alone is unfitted to nature, the skills acquired by Prometheus make artifice—technology—our second nature. As Daniel Bell observes, "The introduction of techne gives man a . . . different character by extending his powers through adaptive skills and redirective thought; it allows him to prefigure or imagine change and then seek to change the reality in accordance with the thought. The fruits of techne create a second world, a technical order which is superimposed on the natural order" (1980, 9). Technology is second nature not only because it extends human powers but also because a skill, once learned, can be practiced intuitively, without reflection. Like the blind man who dwells in his cane, we don't so much use technology as we inhabit and live through it.

This notion of "second nature" broadens our understanding of

1

technology so that it can no longer be fully explained in terms of utility. Contemporary life would scarcely be thinkable without clocks, telephones, or automobiles—not so much because these inventions have become economic necessities (although this too is true) but because we inhabit a world governed by schedules and deadlines, traversed by drivers and pedestrians, and linked by media. But precisely because such inventions have become second nature, we tend to miss their significance—whether aesthetic, economic, or political—in shaping our collective lives. A theory of technology, then, implies a theory of culture.

The story of Prometheus, on this reading, invites two questions. First, the conflict between man's pride and his ingenuity raises the issue of limits: how has technology been guided or directed historically, and what limits, if any, should be placed on its development today? Second, how do institutions mix with technology to evolve new forms of collective life that become second nature?

Returning to Prometheus, Percy Bysshe Shelley much later retold his story, but the ancient hero was recast as a rebel "released from the fetters of gods and freed to do good works" (Winner 1977, 312). Shelley's romantic vision of the technological sublime removed any hint of human weakness or folly. As he stated in the preface of *Prometheus Unbound*, the hero is "as it were, the type of the highest perfection of moral and intellectual nature, impelled by the purest and truest motives to the best and noblest ends" (124).

In what might be described as a memorable domestic quarrel, Mary Shelley challenged her husband's revisionism when she restored the tragic dimension of Aeschylus's work in her own *Frankenstein, or The Modern Prometheus*. Victor Frankenstein, upon mastering the new science of Galileo and Newton, imparts life and intelligence to inert matter only to refuse responsibility for the consequences of his creation. Victor's moral weakness exposes the scientist as an imperfect god who unwittingly causes his own destruction and that of his wretched Adam.

In the figure of Frankenstein's monster, critics today find a convenient literary symbol of technology out of control. And, as in Mary Shelley's Gothic novel, the scientist is often singled out as the culprit whose means, to paraphrase Ahab, are sane but whose ends are mad. Because modern science merges theory and practice, Mary Shelley feared that scientists would refuse responsibility for the consequences of their experiments (Wartofsky).

Where Mary Shelley's Prometheus brings us tragedy, the poet's hero promises romance. Although "romance" usually refers to a literary genre or rhetorical trope, it is also a fair characterization of the En-

2

lightenment "dream of reason" first articulated by Descartes. "Give me extension and motion," he declared, "and I will construct the universe" (Bell 1980, 11). The mind, freed from prejudice, is itself a kind of technology, an instrument of reason that could expand mankind's intellectual and practical powers if only it were applied correctly: "I saw consequently that there must be some general science. This, I perceived, was called universal mathematics. To speak freely, I am convinced that it is a more powerful instrument of knowledge than any other that has been bequeathed to us by human agency, as being the source of all others" (Bell 1980, 11). Of all the possible ends served by his new method, Descartes singled out medicine, the science of healing, as the most worthy. He may have been intoxicated by his vision, but Descartes had no intention of misusing his method—an instrument that does not descend from the gods but originates in man's native reason.

If technology has tragic implications, the source of unhappiness may well reside in those who would block its progress, not in those whose ingenuity furthers its cause. Indeed, the very scholars in the humanities who substitute contemplation for action may be most guilty of the hubris that fated Prometheus in Aeschylus' play. As Martin Kalin points out in this volume, when Descartes' contemporary, Galileo, trained his telescope on the heavens, he was rebuked by papal astronomers who claimed, on the authority of Aristotle, certain knowledge of the number and position of the planets. Is it possible, Kalin asks, that philosophers who now dismiss or abhor the idea of artificial intelligence are the new papal astronomers?

Why, then, would Mary Shelley give her modern Prometheus a tragic profile? Victor Frankenstein offers a clue when, in abandoning the mystical alchemy of Agrippa and Paracelsus, he ruefully admits, "I was required to exchange chimeras of boundless grandeur for realities of little worth" (Wartofsky 179). Descartes' geometric method, it seems, empties the world—now reduced to matter and motion—of human significance.

How we evaluate the Shelleys' respective visions of technological change appears to depend less on evidence than on temperament and aesthetic preference (cf. White 433). If this is the case, then perhaps the lesson of Prometheus has no place in a sober analysis of technological change. This, in fact, is the claim of those who engage in the empirical practice of technology assessment. In their view, criticism of the "technological imperative" is on par with Mary Shelley's tale or, for that matter, Emerson's remark that things are in the saddle and ride mankind. Such admonitions, they say, should be treated not as reasoned argument but as rhetorical protest against offenses to a "tender-minded" sensibility (Mesthene 72–73). Sir Peter Medawar, too, reduces protest to temperament but he also judges it harshly: "There is all the difference in the world between informed and energetic criticism and a drooping despondency that offers no remedy for the abuses it bewails. . . . To deride the hope of progress is the ultimate fatuity, the last word in poverty of spirit and meanness of mind" (125, 127).

Medawar's impatience with protest against technological progress seems justified. After all, artists and humanities scholars belong to a long tradition that laments "mechanistic philosophy" and its outward institutional forms but cannot seem to propose any cure. Jacques Ellul, for example, complains that our current age "scarcely favors spontaneity or creativity." "Nor," he continues, "can it know [the] living rhythms" of an earlier, more pastoral way of life (38) Mary Shelley, like Ellul, might be said to hold simultaneously tragic and romantic beliefs about the place of technology in society. The tragedy of Frankenstein's creation may be attributed to both the power of mechanistic science and the absence of social conscience in those who apply its principles. But Victor also can be seen as a romantic who yearns for an earlier age of contemplation and mystery, an age identified in the novel with pursuit of the "philosopher's stone." In a word, both Mary Shelley and Ellul, admittedly an unlikely couple, look backward for harmony with a natural order, while Percy Bysshe Shelley beckons the future for its promise of nature subdued and man elevated through the unchecked exercise of his powers. Should we dismiss such visions of technology and culture as fanciful or, worse, as humanistic pieties? Is Medawar right when he suggests that technological problems, themselves produced by technology, are best solved by technology as well?

None of the contributors to this volume would endorse Ellul's conception of technology as an oppressive and autonomous force which dominates mankind in every sphere of life. But neither would they dismiss the lessons of Prometheus bound as tender-minded sentiments or allow Percy Bysshe Shelley's Prometheus unbound, however noble his motive, to escape their scrutiny. These writers take discourse about the limits and form of technology seriously.

The Humanities and Public Discourse

Yet the authority of the humanities has been challenged on all sides. The loss of credibility may be due, in part, to the retreat of the humanities from public life. What Christopher Lasch has argued elsewhere about the status of the fine arts in industrial society describes equally well the recent history of the humanities generally understood:

> In industrial societies, art is doubly segregated from everyday life, in the first place because it retains so few of its earlier associations with ritual, sociability, and work, and in the second place because the glorification of art has gone hand in hand with its definition . . . as an activity of the leisure class. When [works of art] lose touch with common experience, they become hermetic and self-referential, obsessed with originality and at the expense of communicability, indifferent to anything beyond the artist's private . . . perception of reality. (Lasch 44)

Historians like Ann Douglas have joined Lasch in arguing that during the Industrial Revolution of the nineteenth century, the humanities retained their purity by accepting a marginal position in the culture. Like a museum piece, their traditions are enshrined but far removed from public life. John Kasson indirectly addresses this phenomenon when he argues that the loss of a vital sense of history is itself a symptom of our culture's obsession with technological progress. Leo Marx also addresses the role of the humanities in public discourse when he notes that the dissenting voices of early American literature and letters—Hawthorne, Melville, Thoreau, Emerson—were barely heard. The diminished stature of this prophetic minority coincided with the emerging belief in progress as the natural consequence of technological expansion.

Much recent discussion in philosophy and social theory centers on the inability of the humanities to locate the grounds of belief in a particular tradition, principle, or ideology (MacIntyre; Bellah et al.). When the public does turn to the humanities for guidance, the academy's confused response cannot inspire much confidence. If the choice between competing ethical principles such as duty and utility cannot be determined rationally, then perhaps the humanities actually contribute to the widespread belief that all commitments are reducible to individual preference. The agreement to disagree—each entitled to his own opinion—substitutes benign acceptance of all private interests for critical inquiry into the shared ends of cultural life. Perhaps Wayne Booth is right when he says that we inhabit the cultural equivalent of Hyde Park, where all are free to proclaim what they please because no one listens (x–xi).

It is fortunate that humanities scholars from many disciplines have articulated genuine alternatives to the relativism and insularity that too often plague academic discourse. Although they do not share the same vision of the ends served by technology—either ideally or in practice—these thinkers enable us to see how political choice and moral commitment permeate all human enterprise. Those who see

5

technology in purely technical terms should note that the cost-benefit calculations of technology assessment carry, whether deliberately or by default, political implications. As Langdon Winner has written, we should be wary of technology—or technology assessment—which is "conspicuously silent about [its] own ends" (1986, 102).

Ian Barbour's *Technology, Environment, and Human Values* demonstrates how technology assessment can be brought within the compass of the humanities. His book groups cultural values under three headings: material values (health, survival, material welfare, employment), social values (distributive justice, participatory freedom, interpersonal community, personal fulfillment), and environmental values (resource sustainability, ecosystem integrity, environmental preservation) (3–5). Barbour shows how such a typology of cultural values can frame analyses of comparable costs, risks, and benefits issuing from alternative technologies. Barbour implies that technology assessment should be guided by criteria broader than technical efficiency or prevention of environmental disasters. Because our culture also should pursue other, sometimes conflicting, aims—participatory freedom, for example—public policy that does not keep a plurality of values in mind is unavoidably myopic and potentially harmful.

Assessments of the breeder reactor, for example, have concentrated on the dangers to the environment or the security risks connected with its production of weapon-grade fuel. Yet they do not investigate the authoritarian managerial structure which such a technology would require of a civilian institution. Compared to the responsibility for managing potentially catastrophic security risks, concern for the civil liberties of workers subject to covert surveillance would seem a mere nicety. Unfortunately, in societies based upon complex technological systems, Winner argues, even raising such political concerns seems foolish before the compelling force of "practical necessity" (1980, 133–34). Is the benefit, to use the language of technology assessment, worth the cost? Democratic values can be protected only if technology assessment moves beyond purely technical considerations and invites the participation of the affected public (Barbour 204).

This volume shows how the choices made in the development of technology, both historically and today, spring from cultural traditions and associated beliefs—among them the Enlightenment faith in the perfectibility of man, the Jeffersonian vision of republican virtue, and the conviction of classical liberalism that natural liberty is a fundamental right. Traditions, then, constitute invisible communities which subtly shape our commerce with the world and our vision of the public good (MacIntyre 175, 178). Even marginal traditions can exert influence. For example, Paul Camenisch's essay examines how the religious beliefs of Jehovah's Witnesses encouraged improvements in medical technology.

Some contributions to this volume examine traditions in conflict, as in Albert Borgmann's discussion of partisan and philosophical differences between liberalism and republicanism. Other essays show how conflict can surface even within a given tradition. For example, Robert Cooke assesses the obligation to employees whose job skills have become obsolete. He examines two philosophical models of rights. Both models arise from the tradition of classical liberalism, yet they yield sharply opposed accounts of rights and obligations.

Although an edited volume cannot fully represent, let alone resolve, complex social issues, we do think this collection is valuable in bringing several traditions to bear upon the place of technology in contemporary America. Historians and philosophers, computer scientists and social scientists, labor leaders and legislators all engage in a lively conversation that weds critical inquiry to the public interest. This collection of essays is a wager that the humanities can speak to public concerns without being exiled to Hyde Park's Speakers Corner.

A Model for Understanding Technology and Culture

Because technology and American culture are inseparable, we need strategies that help us discern their "organized complexity" (Jacobs 428–33). Technology should not be understood on the model of a tool—something consciously chosen to achieve a predetermined end. Unlike a hammer, the use and effect of which is decided by the whim of its user, complex systems and powerful techniques can delimit the field of possibilities for thought and action. We also believe the language of "impacts" and "side effects"—common in literature on these issues—is simplistic in assuming that technology impinges upon society as a thing apart, an efficient cause which produces isolated economic, political, or environmental effects. Technology is not neutral because it embodies the choices made by society, but for the same reason it cannot be treated as an autonomous, impersonal force over which mankind has no control. The truth, that technology and culture are mutually determining, lies somewhere between these two extreme positions.

A less facile, though admittedly more messy, explanation would embed technology in dimensions of culture "all varying simultaneously and in subtly interconnected ways" (Jacobs 433). We have organized the book into four sections, each representing a dimension of collective life: consciousness, work, politics, and community. *Consciousness*, as we use the term, refers to the grounds of belief, as well

7

as the forms of symbolic action, articulated by science, philosophy, religion, ideology, and art. *Work* includes the means and mode of production—machines and organization of labor—as well as the distribution of goods and services that define economic life. *Politics* refers not only to the institutions of government but more generally to the public sphere where citizens contend for power and negotiate collective interests and commitments. Finally, *community* can be understood as self-identity constituted and enlarged by identity with others, its affiliations ranging from families to social movements. Culture is the organic whole constituted by each of these dimensions.

We make no attempt here to elaborate this model, but we do think it preserves our subject's richness and complexity. This model also enables us to pinpoint important questions and issues that focus inquiry. Our categories are deliberately blurred at the edges, because culture as a theme of reflection demands "thick description," not a rigid taxonomy (Geertz 3–30). Christopher Lasch's essay offers a good example of how all four dimensions shade into each other. When he contends that management prizes technology because it ensures control over labor, he immediately draws politics within the domain of the workplace. At the same time, he argues that alternative technologies which promise workers greater control—flexible specialization, for example—must be guided by a revived sense of community, by norms traditionally foreign to American industry. After casting doubt on this prospect, he attributes the manager's struggle for control to a kind of false consciousness, a mistaken but widely shared belief in the power of technology to overcome all human limitations. The focus of his provocative essay is work, but Lasch's rich analysis encompasses politics, community, and consciousness as well. Most essays in this volume can be similarly described. Seen from a particular angle of vision, one dimension stands out sharply, but the others remain in the background and can just as easily focus our attention.

A Thematic Overview

This overview sketches several of the major issues that emerge from this collection of essays and that touch upon our four dimensions of culture. Progress as technological expansion or as the fulfillment of political ideals, mass production or flexible specialization, computer surveillance as the protection or the violation of human rights, economic mobility or cohesive community—these are among the choices engaged by our contributors and by our technological republic.

The conflicting attitudes toward technology embodied by Prometheus surface repeatedly in American history. As John Kasson points out in his essay for this collection, the "unbound" Promethean heroes of our recent past have been inventors like Edison whose electric light bulb had dispelled "night with its darkness . . . from the arena of civilization" (184). Can it be denied that history is the record of progress that reduces labor, enriches leisure, promotes health, and promises abundance? Kasson notes that this is precisely the attitude of Edward Byrn who, in 1900, conducted a frightening thought experiment. Imagine, he said, a journey into the past, a world without "air engines, stem-winding watches, cash registers, the Suez Canal, iron frame buildings . . . [the list goes on for half a page] but enough!" (184–85). Looking back, we, in turn, might take pity on poor Byrn, who somehow lived out his days without the benefit of a word-processor to compose his thoughts. Kasson's reading of Twain's *Connecticut Yankee in King Arthur's Court*, both in this volume and his book *Civilizing the Machine*, exposes the arrogance and danger in Byrn's disdainful attitude toward the past. The story chronicles the effort of Hank Morgan, an enlightened Yankee, to "civilize" feudal Camelot by introducing modern democratic and commercial institutions. To achieve his noble end Hank ultimately must resort to such "liberating" technology as electrified fences and Gatling guns. Helpless before Hank's know-how, thousands are cruelly slaughtered in the name of progress. Twain's novel thus portrays a Prometheus more callous and unheroic than any of the mythic figure's earlier incarnations. Insensitive to genuine historical or cultural differences, Hank personifies arrogance unredeemed by Yankee ingenuity. As with the Prometheus of Mary Shelley, irresponsible actions have tragic consequences—for Hank as well as Arthur's knights.

Perhaps Byrn's technological sublime should not be so enthusiastically embraced, even if, as Orwell wrote, "in practice any attempt to check the development of the machine appears to us as an attack on knowledge and therefore a kind of blasphemy" (Orwell 182). Senator Mathias also alludes to Orwell when he conjures the image of Big Brother as the threat to privacy introduced by computer technology. Computerized data banks and surveillance technology offer great advantages in rooting out fraud, monitoring the movement of criminals, and protecting creditors from financially irresponsible clients. Mathias does not contest that such actions may be desirable, but he believes the new technology, because it is seductive, presents a threat to individual privacy and to democratic principles. Again we see the question of responsible limits, this time deriving from an interpretation of the American Constitution.

As Leo Marx argues in his essay, the idea of limiting technology was not always considered un-American, let alone irreligious. The case can only be decided by what we define as progress. Jefferson, Marx reminds us, feared that domestic manufacturing would create an urban proletariat, would ruin American hopes for a truly republican government. He and many of his contemporaries argued that progress depended upon the adaptation of European technology to an envisioned American "middle landscape" (cf. Marx). This ideal, Jefferson's compass for the march of progress, would find a middle path between a technologically driven civilization and nature's rudeness. But by mid-nineteenth century this vision had succumbed to a rhetoric of the "technological sublime" which extolled the works of man as comparable in power and effect to the works of God. By the end of the nineteenth century faith in human perfectibility was now narrowly identified with the goal of economic prosperity. The belief in progress as a movement away from feudal institutions and toward republican government, Marx continues, was all but forgotten.

Of course, critics of unbounded technology persisted, but their complaints typically were dismissed or ridiculed as nostalgic reveries. To use a mechanical metaphor, technology itself could operate as a kind of thermostat that would regulate the entirety of social and political life. This, of course, was precisely what troubled men like Thomas Carlyle, who did not wish away machinery and mechanistic thinking but feared its intrusion into spheres of life where it did not belong. In his essay "Signs of the Times" (1829) Carlyle declared, "Men are grown mechanical in head and heart, as well as in hand. They have lost faith in individual endeavor, and natural force, of any kind. Not for internal perfection, but for external combinations and arrangements, for institutions, constitutions—for Mechanism of one sort or other, do they hope and struggle. Their whole efforts, attachments, opinions, turn on mechanism, and are of a mechanical character" (Winner 1977, 67). Leo Marx refers to other writers like Emerson, Thoreau, and Melville who, far from being nostalgic primitivists, reconstituted the ideal of the "middle landscape" as an alternative to the dominant belief in technological expansion.

In a surprising reversal of expectations, John Kasson's thoughtful essay suggests that a nostalgic attitude toward the past better characterized our technological heroes, men like Henry Ford and Walt Disney, better than "romantic" writers who might criticize them. Whether Byrn's horror of the past or Twain's nightmarish picture of Hank's attempt to conquer it, visions of history colored by the technological sublime reduce the past to the imperfect present, a primitive stage in the ongoing cycle of material production and consumption. Such an understanding of the past denies the existence of a way of life significantly different from our own. As a consequence, one's historical sense atrophies and any genuine sense of continuity with the past is lost. To recover from this sense of loss, our Fords and Disneys try, by means of advanced technology, to reconstruct the past of their childhood. But such efforts, far from retrieving history, only manage to distort or falsify it.

In this passage from *White Noise*, a recent novel by Don DeLillo, Byrn's thought experiment is repeated but now it is we and not our ancestors who appear primitive:

> We think we're so great and modern. Moon landings, artificial hearts. But what if you were hurled into a time warp and came face to face with the ancient Greeks. . . . What could you tell an ancient Greek that he couldn't say, "Big deal." Could you tell him about the atom? Atom is a Greek word. . . . What good is knowledge if it just floats in the air. It goes from computer to computer. It changes and grows every second of every day. But nobody actually knows anything. (DeLillo 147–49)

Although Tom Edison may have had a thing or two to teach Aristotle, most of us would be denied admission to his Lyceum. Technology guarantees society neither knowledge nor wisdom, although we may deceive ourselves that we either possess or do not need both. Still, it cannot be denied that even if technology breeds passivity in most of us who consume its products without reflection, much that is new is also ingenious and provocative.

Artificial intelligence (AI), the attempt to make computers simulate human intelligence, could, at least in principle, do our thinking for us in much the same way that calculators and spreadsheet packages now do our arithmetic. Rather than confirming the hopes of "Fifth Generation" enthusiasts or the fears of moviegoers who witnessed HAL's nervous breakdown in *2001: A Space Odyssey*, Martin Kalin recommends the modest, analytical attitude of the researcher. Although the philosophical, political, and economic implications of AI research obviously are not technical matters, Kalin cautions that only a clear understanding of how AI works—its powers and limitations—can keep us from indulging one's "temperament" or engaging in idle speculation. Scientists who engineer AI systems have much to teach us about the representation of knowledge and the limitations of rules in understanding apparently simple inferences. In his view, AI can supply a model for intelligence that would not require a biological material, but this does not mean that computers will unseat humans as nature's supremely rational creatures in the near or even distant future. He does find it curious (and disturbing), though, that humanities scholars and social critics would condemn a priori the very possibility of thinking machines. Only time and painstaking research, he concludes, will tell.

Albert Borgmann is concerned less about the metaphysical implications of new technology than its alliance with the political theory and practice of liberalism. For most of us, large-scale, centralized bureaucracies are not really the frighteningly oppressive, authoritarian institutions that Lewis Mumford would have us believe. Quite the opposite, we are grateful for their presence because, without centralization, each of us would require a great deal more knowledge and time to produce the things we desire or need. Confirming the view of the character in DeLillo's novel, Borgmann notes that a highly centralized economy enjoys the advantage of freeing its citizens' time for leisure consumption, an activity that requires little special training or skill. Technology—whether AI or household gadgets—in this view promotes classical liberalism by saving labor and guaranteeing individual freedom to consume mass-produced goods. Elsewhere, Borgmann uses an amusing, if unsettling, example of a Christmas meal to criticize this particular vision of democracy:

> There is a cartoon where a middle-aged woman . . . holding up two packages . . . says to her husband: "For the big day, Harv, which do you want? The traditional American Christmas turkey dinner with mashed potatoes, giblet gravy, oyster dressing, cranberry sauce and tiny peas or the old English Christmas goose dinner with chestnut stuffing, boiled potatoes, brussel sprouts and plum pudding?" The world of bountiful harvests, careful preparations, and festive meals has become a faint and ironical echo. (Borgmann 51)

Borgmann insists that neither appeals for the rejuvenation of community (cf. Bellah et al.) nor the Republican party's assertion of religious and family values can substantively change political life. Such gestures have little effect because the underlying technological order covertly supports the competition and pursuit of individual economic interests. Borgmann does not begrudge economic freedom, but he proposes that contemporary America clear a space for the celebration of community through shared public traditions. Such traditions fall outside the sphere of political parties committed to "repair" of the social order through economic improvement. The celebration of community, by contrast, extends the hope of "redemption" and defines a practical role for government in nurturing civic republicanism, a form of political life transcending the competition of private interests. Apart from such an embodiment, Borgmann believes that communitarian sentiment will evaporate under the glare of "the technological specification of liberalism."

Similarly, in his discussion of high technology and higher education James Carey warns against the narrowly vocational agenda urged by industrial and governmental interests. The university, in his view, should remain an academic community devoted to the intellect and character of the student, not to the production of a commodity for exchange in the marketplace. The traditions of the liberal arts should be protected from the world's daily business so that their worth is never confused with mere utility.

If politics can rise above the marketplace, the opposite is not true. It is a commonplace that Jefferson believed an agrarian economy of yeoman farmers and decentralized, household production would best enact the principles of democracy. Although, for a time the "language of republican industry and virtue melded sweetly into the sound of the factory bell," the reconciliation of Jefferson's pastoral vision and Hamilton's program for a growing industrial economy was short-lived (Kasson 28). As industry grew larger and more centralized, technological innovations offered the means and the pretext for retooling peasants and farmers. Their work ethic of self-reliance and self-determination was upheld in theory but undermined in practice (Gutman; Rodgers). This development was not inevitable but driven by political choices. What confronts us as fate is nothing more than a "hardened history," the convoluted record of roads taken or bypassed (Noble xi; cf. Piore and Sabel 19–48).

Over a decade after Daniel Bell announced the emergence of a post-industrial society (1973), some observers now discern a "second industrial divide" marking the transition to a new stage of industrial production. How we cross that divide, the kinds of technologies that we choose, may well determine the organization and scale of work for the foreseeable future. Again, the issue of how we organize the workplace so that it embodies democratic values has come to the fore.

Drawing upon the work of David F. Noble, Christopher Lasch argues in his essay that at least since Frederick Taylor fathered scientific management in the nineteenth century, new technologies have been selected and implemented for the purpose of consolidating management's power. Against Lasch, it might be said that technological innovation is both welcome and inevitable because new machines and techniques increase efficiency and productivity. But this is precisely the point that Lasch denies. He cites Noble's study of the machine tool industry to support his argument that power, and not efficiency, is at issue (Noble xi–xii, 32, 146). According to Noble, the "record/playback" system for automating machine tools would have given the skilled worker a special role in designing as well as executing the production process. The industry chose instead a numerical control method which places control of production in the hands of engineers. The choice was not between the old and the new but between two new methods distinguished mainly by their political implications (Noble 191–92). Lasch, then, shares with Noble the pessimistic reading that technology tends toward a more centralized and inegalitarian economic order.

As Brian Sullivan points out, other observers of American industry envision a more promising future. Although they differ on the details, Robert Reich, Michael Piore, and Charles Sabel all believe that international economic pressures (especially the growing dominance of Pacific Rim nations in mass production), technological innovation, and the experimental reorganization of work can together bring democracy to the workplace. Automated machinery, wedded to the computer, can lead either to tighter control, as Lasch observes, or to the development of more flexible machinery that would place greater responsibility in the hands of skilled workers (Reich 126–30). Flexible specialization, the second alternative, would return industry to craft modes of production which are egalitarian and cooperative in character.

Lasch greets Piore and Sabel's hopeful scenario with skepticism. Piore and Sabel believe sweeping changes in the workplace would renew the American tradition of yeoman democracy. Lasch, however, believes that this sense of community is necessarily a precondition— and not a consequence—of flexible specialization. Sullivan draws a somewhat different conclusion. Although Piore and Sabel's assessment may prove wrong, he believes that purely economic pressures of global competition will force the United States to return to craft modes of production. Economic interests, in this view, may be sufficient to bring about significant political change within the workplace.

James Carey cautions that if we continue on our present economic course, the "postindustrial" society will also be a two-tiered society. Deskilling of workers, the rampant growth of low-paying jobs in the service sector, and the decline of unionized labor may indeed further polarize the distribution of wealth and power in the United States. Daniel Bell's earlier predictions of a postindustrial society with its meritocratic vision of the rise of a new knowledge class failed to anticipate this dim prospect for the postwar middle class worker (cf. 1973, 128, 143–54).

To avoid this bleak future Irving Bluestone, former vice president of the United Auto Workers, argues the need for broadening the scope of collective bargaining and redefining the social contract that binds management and labor. We must fit technology to the workers rather than the other way around. Although he discerns a trend toward de-industrialization that must be checked, Bluestone does not announce the need for fundamental change in the economic order. Instead, he outlines specific practical steps that management and labor can take, and have taken, to address such issues as job security, compensation, health and safety, and surveillance. Among the more promising efforts noted by Bluestone are the recent UAW pacts with GM and Ford and Quality of Work Life programs which allow employees greater control over the process of production.

Plant closings and their devastating effects on the economy have been well documented. James Fallows has reminded us that abrupt geographical shifts in economic development and decline are the price we pay for a technological-economic order that is "one of the world's most disruptive forces" but also the source of our abundance. "The American bargain is responsible not simply for what is painful in our society but also for what is promising" (Fallows 62–63, 67–68). Weiss and Metzger accept Fallows' premise but they question his conclusion that the flight of industry from a community necessitates its decay. Local economies can be revitalized, they believe, and community organizations can play an important role. Taking the recent history of Pittsburgh as their case study, Weiss and Metzger show how collective bargaining can be used effectively by communities in negotiating technological development with local government, education, and industry. Their study traces a cooperative planning effort that determined the eventual use of an abandoned steel mill and led to the formation of a sophisticated community development corporation. Such efforts go much further than many partnerships between government and industry which tend to create "dumbbell economies" where stagnation spreads between flourishing corporate headquarters and financial centers at the core of urban areas and clusters of high tech industries along the beltways. Two questions should guide community efforts: Who decides and what will be the consequences for the surrounding neighborhood? Weiss and Metzger believe the Pittsburgh case teaches valuable lessons that might help other communities answer these same questions.

Robert Rotenberg's comparative study of Austrian and American responses to the organization of work begins with a simple but important observation. The number and distribution of hours in a work week determines our time outside work. If our family commitments or leisure pursuits suffer as a consequence, then we should rethink our expenditure of time in the work place. The Austrians have done just that; we have not. The contrast, Rotenberg argues, may be understood in terms of the political history of the two countries. Austria has

15

a viable socialist party and a strong organization of labor. Austrians insist that their life outside work shape their life within the work place while Americans permit work to define their social lives without remainder.

Robert Cooke offers a different perspective in this discussion of technological change in the workplace. Unlike the other contributors who recognize that the social good cannot be translated into economic self-interest, he argues that the elimination or redistribution of jobs due to obsolete skills places no special obligation on society or the employer for retraining or job placement. According to Cooke, the latest round of technological innovation raises no unprecedented problem and requires no special solution. His essay examines two philosophical models of justice—the Natural Liberty Model of Nozick and the Social Good Model of Rawls—and concludes that the principle of individual free choice—both the employer's and the employee's—must define rights and corresponding obligations in a free society. As a consequence, employers are morally obligated to place or retrain employees with obsolete skills only if a prior agreement to this effect has been reached by both parties.

Paul Camenisch and Albert Borgmann move from utilitarian and contractual concerns to emphasize the intrinsic importance of carving out a space for communal life in our technological society. We have already mentioned Borgmann's vision of "communities of celebration." In his discussion of medical technology, Camenisch turns away from government and labor as agents of revitalization to see how religious communities, voluntary associations, and the professions can play a mediating role in determining the shape and limits of technological development. Unlike Borgmann, who believes communities begin where technology ends, Camenisch argues that subsidiary communities have achieved modest success in reconciling institutional practices with various conceptions of healing. For example, he maintains that the Jehovah's Witnesses' prohibition against transfusions actually contributed to the development of techniques for bloodless surgery; and he indicates that the concept of hospice care, originally conceived as an alternative to hospital technology, has influenced how hospitals now care for the terminally ill. The artificial heart, Camenisch points out, is an undecided test case which forces physicians to reconcile their desire to improve medical technology with their commitment to the well-being of the patient.

The efforts of these small, diverse groups investigated by Camenisch recall a striking image from Daniel Boorstin's Homeric epic of America's early national experience. Boorstin retells the story of a wagon train called to a halt only a few days out of Independence, Missouri,

16

so the company could form a constitution and establish itself as a mobile polis. "The nation," comments Boorstin, "was beginning not at
one time or place, but again and again, under men's very eyes. Americans were forming new communities and reforming old communities
all over the wild expanse of the Western world" (1965, v). In a later
work, however, Boorstin now places faith solely in technology as the
agent of democracy which removes the barriers of race, class and religion. The automobile, with its expressways and commercial strips,
represents the spread of a homogeneous, global culture, a "Republic
of Technology" (1978, xiv–xv, 2–3, 9–10) that will not pause to create
communal structures to carry us into the future (1978, 9, 24–25).
Camenisch and Borgmann return us to the American tradition praised
by Boorstin in his earlier work. As an alternative to the undirected expansion of technology, they emphasize the power and will of human
beings to harness technology and to shape diverse centers of common life.

Works Cited

AESCHYLUS
1953   *Prometheus Bound*. Translated by Edith Hamiliton. In *Greek
                Plays in Modern Translation*. Edited by Dudley Fitts. New
                York: Basic Books.

BARBOUR, IAN
1980   *Technology, Environment, and Human Values*. New York:
                Praeger.

BELL, DANIEL
1973   *The Coming of the Post-Industrial Society*. New York: Basic
                Books.
1980   *The Winding Passage*. Cambridge, Mass.: ABT Books.

BELLAH, ROBERT, RICHARD MADSEN, WILLIAM M. SULLIVAN, ANN SWIDLER,
and STEVEN M. TIPTON
1985   *Habits of the Heart: Individualism and Commitment in American
                Life*. Berkeley: University of California Press.

BORGMANN, ALBERT
1984   *Technology and the Character of Contemporary Life*. Chicago:
                University of Chicago Press.

BOORSTIN, DANIEL
1965   *The Americans: The National Experience*. New York: Vintage
                Books.
1978   *The Republic of Technology*. New York: Harper and Row.

DELILLO, DON
1985   *White Noise*. New York: Vintage Books.

DOUGLAS, ANN
1977   *The Feminization of American Culture*. New York: Avon Books.

17

ELLUL, JACQUES
1980   *The Technological System*. Translated by Joachim Neuysro-
             schel. New York: Continuum Books.
FALLOWS, JAMES
1985   "America's Changing Economic Landscape," *The Atlantic
             Monthly* 251: 47–68.
GEERTZ, CLIFFORD
1973   *The Interpretation of Cultures*. New York: Basic Books.
GUTMAN, HERBERT G.
1976   *Work, Culture and Society in Industrializing America*. New
             York: Vintage Books.
JACOBS, JANE
1961   *The Death and Life of Great American Cities*. New York: Vintage
             Books.
KASSON, JOHN F.
1976   *Civilizing the Machine: Technology and Republican Values in
             America, 1776–1900*. New York: Penguin Books.
LASCH, CHRISTOPHER
1984   "The Degradation of Work and the Apotheosis of Art,"
             Harper's 268: 40–45.
MACINTYRE, ALASDAIR
1981   *After Virtue: A Study in Moral Theory*. Notre Dame, Ind.:
             University of Notre Dame Press.
MARX, LEO
1964   *The Machine in the Garden: Technology and the Pastoral Ideal in
             America*. New York: Oxford University Press.
MEDAWAR, SIR PETER
1972   *The Hope of Progress*. London: Methuen Press.
MESTHENE, EMMANUEL G.
1984   "Technology as Evil: Fear or Lamentation?" *Research in Phi-
             losophy and Technology* 7: 59–74.
NOBLE, DAVID F.
1986   *Forces of Production*. New York: Oxford University Press.
ORWELL, GEORGE
1982   *The Road to Wigan Pier*. New York: Penguin.
PIORE, MICHAEL J., and CHARLES F. SABEL
1984   *The Second Industrial Divide*. New York: Basic Books.
REICH, ROBERT B.
1983   *The Next American Frontier*. New York: Times Books.
RODGERS, DANIEL
1978   *The Work Ethic in Industrial America, 1850–1920*. Chicago:
             University of Chicago Press.
SHELLEY, PERCY BYSSHE
1966   *Prometheus Unbound*. In *The Selected Poetry and Prose of Shelley*.
             Edited by Harold Bloom. New York: Meridian Press.
WARTOFSKY, MARX W.
1984   "Good Science—Bad Science or Dr. Frankenstein's Di-
             lemma," *Research in Philosophy and Technology* 7: 177–90.

PART ONE

Consciousness

1

Does Improved Technology Mean Progress?

LEO MARX

If some variant of this question had been addressed to a reliable sample of the American people at any time since the early nineteenth century, the answer of a large majority would have been an unequivocal yes. That is the view currently held, I believe, by most historians. The idea that technological improvements are a primary basis for— and an accurate gauge of—the general *progress* of society has long been a fundamental article of belief in the United States. In the last half century, however, that belief manifestly has lost some of its credibility, and a growing minority of Americans has adopted a skeptical, even negative, view of technological innovation as an index of social progress.

The extent of this change in American attitudes was brought home to me recently when I spent a month (October 1984) in China as a member of a delegation of scholars surveying the state of American studies in Chinese universities. At that time the announced goal of the People's Republic was to carry out (in the popular slogan) "Four Modernizations"—the four being agriculture, industry, science and technology, and the military. What particularly struck our group of Americanists was the seemingly unbounded, exuberant, largely uncritical ardor with which the Chinese were conducting their love affair with that "foreign" or "Western" style of modernization—individualistic, entrepreneurial, or "capitalist," as well as scientific and technological—which formed the historical ground for the American ideal of progress. Our delegation's reaction to the Chinese enthusiasm for our kind of modernity reminded me of those astonished European observers who had encountered a similar frenzy of innovation or "modernization" in the United States in the period, say, between 1830 and 1860 (Fisher). Like those European travelers, we were witnessing a society in a veritable transport of improvement; long pent-up innovative energies were being released, everyone seemed to be in motion, everything was eligible for change, and it was assumed that any such change almost certainly would be a change for the better.

As it happens, most of the Chinese we came to know best—teachers and students of American studies—explicitly associated the kind of progress represented by the "four modernizations" with the United States. We talked to large, enthusiastic audiences of university students and faculty about American society and culture, and many, perhaps most, seemed to see this country through a lens colored by their progressive aspirations. The United States, after all, had already achieved many of the objectives China was setting for itself. Their respect for American wealth and power was flattering but disconcerting, for we often found ourselves reminding the Chinese of serious shortcomings, even some terrible dangers, inherent in the Western mode of industrial development. Like the Americans whom the European travelers met 150 years ago, many of the Chinese seemed to us to be extravagantly, excessively, almost blindly, credulous and optimistic. What our reaction had revealed, among other things, was a change in our own culture and, in some cases, in our own personal attitudes. We had been brought face to face with the gulf that separates the outlook of many contemporary Americans from the old national faith in the advance of science-based technology as the basis of social progress.

How shall we explain this change in attitude? A standard answer is to invoke that familiar litany of death and destruction which distinguishes the recent history of the West: the two barbaric world wars, the Nazi holocaust, the Stalinist terror, and the nuclear arms race. It is striking to note how many of the fearful events of our time involve the destructive use or misuse, the unforeseen consequences, or the disastrous malfunction, of modern technologies: Hiroshima and the nuclear threat; pollution and other kinds of damage inflicted upon the environment by advanced industrial societies; spectacular accidents like Three Mile Island, Bhopal, the explosion of the space shuttle *Challenger*, and the Chernobyl nuclear calamity. Such conspicuous disasters admittedly have done a great deal to undermine the public's faith in progress, but there is a longer term change in our thinking to which I, as a cultural historian, want to call attention. It is less ob-

24

vious, less dramatic and tangible than the record of catastrophe that distinguishes our twentieth-century history, but I believe it is more fundamental. I want to suggest that our very conception—our chief criterion—of progress has undergone a subtle but decisive change since the founding of the Republic, and that change is at once a cause and a reflection of our current disenchantment with the old progressive ideal.

To chart this change in attitude we need to go back at least as far as the first Industrial Revolution. The development and diffusion of the radically improved machinery (based on mechanized motive power) used in the new factory system of the late eighteenth century provided what was widely taken to be tangible confirmation of the high hopes aroused by the earlier triumphs of the natural sciences. The emergence of these new technologies coincided with the formulation and diffusion of the modern Enlightenment idea of history as a record of progress. This conception was the fulcrum of what was to become the dominant American worldview. As developed by writers like Condorcet and Turgot, Paine and Priestley, Franklin and Jefferson, the progressive worldview rested on the assumption that history, or at least modern history, is driven by the steady, continuous, cumulative and, indeed—for the truest of true believers—inevitable (or preordained) expansion of human knowledge of, and power over, nature. This new scientific knowledge and technological power was expected to make possible a comprehensive improvement in all the conditions of life—social, political, moral, intellectual, as well as material. Thus, history was conceived as a record of ongoing progress.

The modern idea of progress emerged in an era of political revolution, and as invoked by its radical French, English, and American adherents, it was considered a revolutionary doctrine in both a negative (or critical) and positive sense. The idea of progress was indissolubly bonded to the idea of a radical struggle for freedom from feudal forms of domination. To ardent republicans like Condorcet, Priestley, or Franklin, a necessary criterion of progress was the achievement of political and social liberation. They regarded the new sciences and technologies not as ends in themselves, but as instruments for carrying out a comprehensive transformation of society. The new knowledge and power would provide the basis for alternatives to the deeply entrenched authoritarian, hierarchical institutions of *l'ancien regime*: monarchical, aristocratic, and ecclesiastical. Thus Jefferson describes the combined effect of the new science and the American revolution upon the minds of Europeans:

But even in Europe a change has sensibly taken place in the mind of man. Science had liberated the ideas of those who read and reflect, and the American example had kindled feelings of right in the people. An *insurrection* has consequently begun, of science, talents, and courage, against rank and birth, which have fallen into contempt. . . . Science is progressive. (Letter to John Adams, October 28, 1813, Cappon 2:391; emphasis mine)

Many of Jefferson's contemporaries shared his view of the dual function of science and technology (1) as instruments of emancipation from ignorance, superstition, and political oppression and (2) as basic resources in the building of a more productive, humane, and just society (Wolin 9–12).

In Jefferson's time, then, the progressive attitude toward science and technology was bound up with the generous and hopeful, if problematically unbounded, political aspirations of the Enlightenment. Admittedly, the idea of history as endless progress did encourage extravagantly optimistic expectations, and, in its most extreme form, indeed, it fostered some wildly improbable dreams of the "perfectibility of Man" and of humanity's absolute mastery of nature. Nor can there be much doubt that the Faustian aspiration toward total control discussed by Christopher Lasch was closely interwoven with this same complex of ideas.[1] Yet the express political beliefs of the radical republican theorists of the eighteenth century, such as the principle of making the authority of government dependent upon the consent of the governed, often had the effect of limiting or circumscribing those aspirations to omnipotence.

The constraining effect of such ultimate, long-term political goals makes itself felt, for example, in Thomas Jefferson's initial reaction to the prospect of introducing the new manufacturing system to America. Although a committed believer in the benefits to be gained from advances in science and technology, he rejected the idea of developing an American factory system ("Let our work-shops remain in Europe," he wrote) on the ground that the emergence of an urban proletariat, which he then regarded as an inescapable consequence of a European-style factory system, would be too high a price to pay for any potential improvement in the material standard of living. He regarded the existence of manufacturing cities and an industrial working class as incompatible with republican government or the happiness of the

1. See below, chapter 5, pp. 87–88. The issue raised here, namely, the extent to which the Enlightenment gave rise to a fundamentally dangerous program of domination by instrumental rationality, has been the subject of a rich debate in German social theory. It centers upon the work of Max Weber and his followers and critics of the "Frankfurt School" (Weber; Adorno and Horkheimer; Habermas).

people. He argued that it was preferable, even if more costly in strictly economic terms, to ship raw materials to Europe and import manufactured goods. "The loss by the transportation of commodities across the Atlantic will be made up in happiness and permanence of government" (Jefferson, Query xix).[2] In weighing relatively intangible political, moral, and aesthetic costs against economic benefits, he anticipated the viewpoint of the environmentalists of our time.

Another instance of the constraining effect of republican political ideals upon self-interest is Benjamin Franklin's refusal to exploit his inventions as a source of private profit. Thus Franklin's reaction when the Governor of Pennsylvania urged him to accept a patent for his successful design of the "Franklin stove":

> Governor Thomas was so pleased with the construction of this stove as described in . . . [the pamphlet] that . . . he offered to give me a patent for the sole vending of them for a term of years; but I declined it from a principle which has ever weighed with me on such occasions, namely; viz., *that as we enjoy great advantages from the inventions of others, we should be glad of an opportunity to serve others by any invention of ours, and this we should do freely and generously.* (108; emphasis in original)

What makes the example of Franklin particularly interesting is the fact that he later came to be regarded as the archetypal, self-made American and the embodiment of the Protestant work ethic. (When Max Weber sought out of all the world *the* exemplar of that mentality for his seminal study, the *Protestant Ethic and the Spirit of Capitalism*, whom did he choose but our own Uncle Ben?) But Franklin's was a principled and limited self-interest. In his *Autobiography*, he told the story of his rise in the world not to exemplify a merely personal success, but rather to illustrate the achievements of a "rising people." He belonged to that heroic revolutionary phase in the history of the bourgeoisie when that class saw itself as the vanguard of humanity and its principles as universal principles. He thought of his inventions as designed, not for his private benefit, but for the benefit of all.

With the further development of industrial capitalism, however, a quite different conception of technological progress gradually came to the fore in the United States. Americans celebrated the advance of science and technology with increasing fervor, but now they detached the idea from the goal of social and political liberation. For one thing, many regarded the eventual attainment of that goal as having been as-

2. For a longer analysis of Jefferson's reaction to industrialization as an expression of his commitment to a pastoral view of life, see Leo Marx (1964, 116–50).

sured by the victorious Revolution and the founding of the Republic. The difference between this later viewpoint and that of Jefferson's and Franklin's generation can be heard in the characteristic rhetoric of Daniel Webster. He and Edward Everett were perhaps the leading public communicators of this new version of the progressive ideology. When Webster decided to become a Senator from Massachusetts instead of New Hampshire, the change was widely interpreted to mean that he had become the quasi-official spokesman for the new industrial manufacturing interest. Thus Webster, who was generally considered to be the nation's foremost orator, was an obvious choice as speaker at the dedication of new railroads. Here is a characteristic peroration of one such performance in 1847:

> It is an extraordinary era in which we live. It is altogether new. The world has seen nothing like it before. I will not pretend, no one can pretend, to discern the end; but everybody knows that the age is remarkable for scientific research into the heavens, the earth, and what is beneath the earth; and perhaps more remarkable still for the application of this scientific research to the pursuits of life. The ancients saw nothing like it. The moderns have seen nothing like it till the present generation. . . . We see the ocean navigated and the solid land traversed by steam power, and intelligence communicated by electricity. Truly this is almost a miraculous era. What is before us no one can say, what is upon us no one can hardly realize. The progress of the age has almost outstripped human belief; the future is known only to Omniscience. (Quoted in Marx 1964, 214)[3]

By the 1840s, as Webster's rhetoric suggests, the process of dissociating the idea of progress from the Enlightenment vision of political liberation was well advanced. Elsewhere in the speech, to be sure, Webster makes the obligatory Whig bow to the democratic import of technological change, but it is clear that he is casting the new machine power as such, quite apart from its political significance, as *the* prime exemplar of the overall progress of the age. He invests the railroad with a quasireligious inevitability that lends force to the characterization of his language as the "rhetoric of the technological sublime." (The power and majesty of the new works of man are comparable, in their effect upon mind—in their capacity to evoke a sense of sublimity—to the works of the Creator.) Speaking for the business and industrial elite, Webster and Everett thus depict science-based tech-

---

3. For a more complete analysis of the passage in the context of the tension between the pastoral and progressive worldviews, see Marx (1964, 209ff.).

nological innovations (and the material improvements they made possible) as a sufficient cause, *in itself*, for the fact that history is assuming the character of continuous, cumulative, overall progress.

At the same time, the dropping away of the radical political ideals of the Enlightenment allowed for the blending of the idea of technological progress with certain other grandiose national aspirations. It is evident that Webster's version of the "rhetoric of the technological sublime" is of a piece with the soaring imperial ambitions figured forth by the slogan "Manifest Destiny" and by such tacit military figurations of American development as the popular notion of the "conquest of nature" (including the Native Americans) by the increasingly technologized forces of advancing European-American "civilization." These future-oriented themes were easily harmonized with the millenial strain in evangelical Protestantism, the most popular American religious dispensation at the time. Webster indicates as much when, at the end of his tribute to the new railroad, he glibly brings in "Omniscience" as the ultimate locus of the meaning of progress.

The difference between the earlier Enlightenment conception of progress and that exemplified by Webster is in large measure attributable to the difference in the groups they represented. Thus Franklin, Jefferson, and the heroic generation of founding revolutionists constituted a distinct, rather unusual social class. It was unusual, in America at least, because for a short time the same men possessed authority and power in most of its important forms: economic, social, political, and intellectual. When we leap ahead two generations, however, we find that men of a very different stripe—the industrial capitalists for whom Daniel Webster was a spokesman—were rising to dominance in American society. They derived their status from a different kind of wealth and power, and their conception of progress, like their economic and social aspirations, was correspondingly different. The new technology and the immense profits it generated belonged to them, and since they had every reason to assume that they would retain their property and power, they had a vested interest in technological innovation. It is not surprising, under the circumstances, that as industrialization proceeded they became true believers in technological improvement as the primary basis for—as virtually tantamount to—universal progress.

In retrospect, and simplifying greatly, the dissociation of technological and material advancement from the larger political vision of progress may be seen as an intermediate stage in the eventual impoverishment of that radical eighteenth century progressive worldview. (Webster, whose notorious compromises with the Southern slave power shocked New England intellectuals like Emerson, may be said to exemplify the detachment of the idea of progress from its political ties with principled republicanism.) This subtle change prepared the way for the emergence, later in the century, of a thoroughly technocratic idea of progress. It may be called "technocratic" in that it valued improvements in power, efficiency, and rationality as sufficient ends in themselves. Among those who bore witness to the widespread diffusion of this concept in the final decades of the century were Andrew Carnegie, Henry Adams, Thomas Edison, Thorstein Veblen, and Frederick Winslow Taylor and his followers. Taylor's theory of scientific management embodies the quintessence of the technocratic mentality, "the idea that human activity could be measured, analyzed, and controlled by techniques analogous to those that had proved successful when applied to physical objects" (Aitken 16).[4]

The technocratic idea of progress is a belief in the sufficiency of scientific and technological innovation as the basis for general progress. It says that if we can ensure the advance of science-based technologies or, in a word, if we can perfect the means—the material base—the rest will take care of itself. (The "rest" here refers to nothing less than a corresponding degree of improvement in the social, political, and cultural conditions of life.) On this view, instrumental values are fundamental to the progress of society, and what formerly were considered to be primary, goal-setting values (justice, freedom, harmony, beauty, or self-fulfillment) may now be relegated to a secondary or epiphenomenal status.

But the old republican vision of progress—the vision of advancing knowledge empowering humankind to establish a less hierarchical, a more just and peaceful society—did not disappear. If it no longer inspired Webster and his associates, it lived on in the minds of many farmers, artisans, factory workers, shopkeepers, small businessmen, as well as professionals, artists, intellectuals, and other members of the lower middle and middle classes. During the late nineteenth century, a number of disaffected intellectuals sought new forms for the old progressive faith. They translated it into such special political idioms as utopian socialism, the single tax, populism, progressivism, and Marxism and its native variants. But in this century such latter-day revisions of the old idealistic Enlightenment belief in social progress have been accorded a peripheral place in American culture. So has the relatively small-scale, personal, interactive technological style (the mode of technological practice itself) that accompanied the ear-

4. The three foregoing paragraphs are a slightly revised version of a passage from my previous essay (1984, 638–52).

lier, artisanal phase of industrialization. That style was supplanted, around the beginning of this century, by Fordism and an obsessive interest in economies of scale, standardization of process and product, and control of the workplace. This shift to the mass production style was accompanied by the triumph of Hamiltonian principles, and by the more or less official commitment of the United States government to the growth of the nation's wealth, productivity, and global power, and to the most rapid possible rate of technological innovation, as the essential criteria, the necessary and sufficient conditions, of social progress.

Let me turn back to the late eighteenth century and pick up another set of ideas which has proven to be the basis for a powerful critique of the culture of advanced industrial society. Usually described as the viewpoint of the "counter-Enlightenment" or the "romantic reaction," these ideas may be said in retrospect to have been the basis for a surprisingly long-lived adversary subculture. According to the conventional wisdom of historians, they had their origin in the intellectual backlash from the triumph of the natural sciences we associate with the great discoveries of Galileo, Kepler, Harvey, and Newton. Put differently, this critical tendency was a reaction against the extravagant claims for the universal, not to say exclusive, truth value of the version of scientific rationalism widely known at the time as "the Mechanical Philosophy." That term derived from the ubiquity of the machine metaphor in the work of Newton and other natural scientists ("celestial mechanics") and many of their philosophic allies, notably Descartes, all of whom tended to conceive of nature itself as a "great engine" and its subordinate parts (including the human body) as lesser machines.

By the late eighteenth century, a powerful set of critical, antimechanistic ideas was being developed by Kant, Fichte, and other German idealists, and by great English poets like Coleridge and Wordsworth. But in their time the image of the machine also was being invested with greater tangibility and social import. The Industrial Revolution was gaining momentum, and as power machinery was more widely diffused in Great Britain, Western Europe and North America, the machine acquired much greater resonance: it came to represent *both* the new technologies based on mechanized motive power and the mechanistic mind-set of scientific rationalism. Thus Thomas Carlyle, who had been deeply influenced by the new German philosophy, announced in his seminal 1829 essay, "Signs of the Times," that the right name for the dawning era was the "Age of Machinery" (3:5–30; Marx 1964, 170–90). It was to be the Age of Machinery, he warned, in every "inward" and "outward" sense of the word, meaning that it would be dominated by mechanical (instrumental or utilitarian) thinking as well as by actual machines—by machine technologies. In his criticism of this new era, Carlyle took the view that neither kind of "machinery"— the rationalistic mode of thought or the new machine-powered technology—was inherently dangerous. In his opinion, indeed, they represented potential progress as long as neither was allowed to become the exclusive or predominant mode in its respective realm. What was needed, he argued, was the proper "coordination," or balance, between the mechanical and what he called "the dynamical." In the realm of thought the "dynamical" stood for those inescapably subjective, intuitive, inexact procedures that were called forth by the interpretation of experience involving human expression, thought, and behavior—procedures which seemed to be excluded or diminished by the new scientific rationalism.

In the United States this ideological tendency was adopted by a small, gifted, if disaffected minority of writers, artists, and intellectuals. Their native version of Carlyle's critical viewpoint also was labeled "romantic" (or "transcendentalist"), with the dismissive implication that it was an excessively idealistic, nostalgic, or sentimental, hence impractical and unreliable viewpoint. Their ideas were particularly vulnerable to that charge at a time when the rapid improvement in the material conditions of life in the United States was lending a compelling power and plausibility to the idea of history as a record of universal progress. It is only in the late twentieth century, indeed, with the growth of skepticism about scientific and technological progress, and with the consequent emergence of a vigorous adversary culture in the 1960s, that the standpoint of that earlier eccentric minority has been accorded a certain intellectual respect.

In the work of writers like Emerson and Thoreau, Hawthorne and Melville, we encounter critical responses to the onset of industrialism which cannot be written off as the expression of mere nostalgia or primitivism. These writers were not primitivists. They did not hold up an idealized wilderness, or a pre-industrial Eden, as preferable to the world they saw in the making. Nor did they dismiss the worth of material improvement as such. But they did regard the dominant progressive belief system, often represented (as in Webster's speech) by the appearance of the new machine power in the American landscape, as dangerously shallow, materialistic, and one-sided. Fear of "mechanism," in the several senses of that word—especially the domination of the individual by impersonal systems—colored all of their thought. In their work, accordingly, the complex bipolar image of the-machine-in-the-landscape, far from being an occasion for exultation, often seems to arouse feelings of anxiety, dislocation, and foreboding. I discuss many of these images in *The Machine in the Garden* (1964), and won't repeat that discussion here. But I do want to emphasize the serious criticism of the debased, or technocratic, concept of progress implicit in that body of imagery. Henry Thoreau's detailed, carefully composed account of the intrusion of the railroad into the Concord woods is a good example; it bears out his witty and trenchant delineation of the new inventions as "improved means to unimproved ends" (46).

This critical view of the relationship between technological means and social ends did not merely issue in random images, phrases, or narrative episodes. Indeed, the whole of *Walden* may be read as a sustained attack on a culture which has allowed itself to become confused about the relationship of ends and means. As a result, Thoreau's countrymen are being depicted as becoming "the tools of their tools" (33). Much the same argument underlies Hawthorne's satire "The Celestial Railroad," a modern replay of *Pilgrim's Progress* in which the hero, Christian, realizes too late that his comfortable railroad journey to salvation is taking him to hell, not heaven. Melville incorporates a similar insight into his characterization of Captain Ahab, who is the embodiment of the Faustian aspiration toward domination and total control to which the sudden emergence of exciting new technological capacities lent a certain novel credence. He exults in his power over the crew, and he explicitly identifies it with the power exhibited by the new railroad in spanning the North American continent. In reflective moments, however, he also acknowledges the self-destructive nature of his own behavior: "Now in his heart, Ahab had some glimpse of this, namely, all my means are sane, my motive and my object mad" (Ch. 41, 161).

But of course there was nothing new about the moral posture adopted by these American writers. Indeed, their attitude toward the largely uncritical, exuberant national celebration of the railroad and other new inventions no doubt is traceable to traditional moral and religious objections to such an exaggeration of human powers. In this view, the worshipful attitude of Americans toward these new instruments of power had to be recognized for what it was: idolatry like that attacked by Old Testament prophets, in a disguised, new-fashioned, characteristically modern form.

This moral critique of the debased or technocratic version of the progressive worldview has slowly gained adherents since the mid-nineteenth century, and by now it is one of the chief ideological supports of an adversary subculture in the United States. To be sure, it is chiefly the viewpoint of a relatively small minority of discontented

LEO MARX

members of the white middle class, but there have been times, like the
Vietnam upheaval of the 1960s, when that minority has won the temporary support of, or formed a tacit coalition with, a remarkably large
number of other disaffected Americans. Much the same antitechno-
cratic viewpoint has made itself felt in various dissident movements
and intellectual tendencies since the 1960s: the antinuclear movements (both against nuclear power and nuclear weaponry); some
branches of the environmentalist and feminist movements; the voluntary simplicity movement; the "small is beautiful" and "stable state"
economic theories, as well as the quest for "soft energy paths" and for
"alternative (or appropriate) technologies."

Perhaps this oversimplified historical summary will help to explain
that strange ambivalence toward the ideal of progress expressed by
many Americans nowadays. As compared with prevailing attitudes in
the United States in the 1840s, when the American situation was more
like that of China today, the current mood in this country would have
to be described as mildly disillusioned. To appreciate the reasons for
that disillusionment, let me repeat, it is necessary to be clear about the
distinction between the two views of progress on which this analysis
rests. The initial Enlightenment belief in progress turned upon the
idea of history as the record of continuously and cumulatively expanding knowledge and power. That growing power, as manifested
by advances in science and technology, was perceived to be in the service of liberation from political oppression. The ultimate confirmation
of this idea of progress was to be a sociopolitical result: the emergence
of a more just republican society, one in which the gap between the
powerful and the powerless, the rich and the poor, would be significantly narrower than hitherto.

But over time that conception of progress was transformed, or
partly supplanted, by another, essentially technocratic, version of the
idea. This is the now familiar view that innovations in science-based
technologies are in themselves a sufficient and reliable basis for progress. The distinction, then, turns on the apparent loss of interest in,
or unwillingness to name, the societal ends for which the scientific
and technological instruments of power previously had been regarded,
in the Enlightenment version, as the means. What we seem to have
instead of a guiding political goal, as Quentin Skinner put it somewhere, is a new minimalist definition of civic obligation.

The distinction between two versions of the belief in progress is
helpful, I think, in sorting out reactions to the many troubling issues
raised by the diffusion of high technology. When, for example, the in-

34

troduction of some new automated or labor-saving technology is proposed, it is useful to begin making a judgment of its worth by calling into question the familiar assumption that the innovation necessarily, or self-evidently, represents a desirable form of "progress." What is the purpose of this new technology? The aim may well be to reduce labor costs, yet in our society the human costs to, say, the displaced workers—the people most directly involved—is likely to be ignored. The same essential defect of this mind-set also becomes evident when the president of the United States calls upon those who devised nuclear weapons to provide an elaborate new system of weaponry, the Strategic Defense Initiative, as the only reliable means of avoiding nuclear war. Not only does he invite us to put all of our hope in a "technological fix," but he combines that argument with a rejection of the ordinary, less perfectionist, but finally indispensable method of international negotiation and compromise. Here again, as so often with this curious viewpoint, technology is thought to obviate the need for political ideas and practices.

One final word. I perhaps need to clarify the claim that it is the modern, technocratic version of the progressive worldview, not the Enlightenment view, which encourages the more dangerous contemporary fantasies of domination and total control. The point is that the political and social aspirations of the generation of Benjamin Franklin and Thomas Jefferson *provided tacit limits to, as well as ends for, the progressive vision of the future*. But the technocratic version so popular today entails a belief in the worth of scientific and technological innovations as ends in themselves. All of which is to say that we urgently need a set of political, social, and cultural goals comparable to those formulated at the beginning of the industrial era if we are to make accurate assessments of the worth of new technologies. Only such goals can provide the criteria required to make rational and humane choices among alternative technologies and, more important, among alternative long-term policies.

Does improved technology mean progress? Yes, it certainly could mean just that. But only if we are willing and able to answer the next question: progress toward what? What is it that we want our new technologies to accomplish? What do we want beyond such immediate, limited goals as achieving cost efficiencies or eliminating the troubling human element from our workplaces? In the absence of answers to these questions, technological improvements may very well turn out to be incompatible with genuine, that is to say, *social* progress.

Works Cited

AITKEN, HUGH
1960   *Taylorism at Watertown Arsenal: Scientific Management in Action, 1908–15.* Cambridge: Harvard University Press.

ADORNO, THEODOR, and MAX HORKHEIMER
1972   *The Dialectic of Enlightenment.* New York: Herder and Herder.

CAPPON, LESTOR J., ED.
1959   *The Adams-Jefferson Letters.* 2 vols. Chapel Hill: University of North Carolina Press.

CARLYLE, THOMAS
n.d.   *Critical and Miscellaneous Essays.* 4 vols. New York: Belford, Clarke and Co.

FISHER, MARVIN
1967   *Workshops in the Wilderness: The European Response to American Industrialization, 1830–1860.* New York: Oxford University Press.

FRANKLIN, BENJAMIN
1958   *Autobiography and other Writings.* Edited by Russel B. Nye. Boston: Houghton Mifflin.

HABERMAS, JÜRGEN
1971   *Knowledge and Human Interests.* Translated by Jeremy Shapiro. Boston: Beacon Press.

JEFFERSON, THOMAS
1955   *Notes on the State of Virginia.* Edited by William Peden. Chapel Hill: University of North Carolina Press.

MARX, LEO
1964   *The Machine in the Garden: Technology and the Pastoral Ideal in America.* New York: Oxford University Press.
1984   "On Heidegger's Conception of 'Technology' and its Historical Validity," *The Massachusetts Review* 25: 638–52.

MELVILLE, HERMAN
1967   *Moby-Dick.* New York: Norton Critical Edition.

THOREAU, HENRY DAVID
1950   *Walden and Other Writings.* New York: Modern Library College Edition.

WEBER, MAX
1958   *The Protestant Ethic and the Spirit of Capitalism.* Translated by Talcott Parsons. New York: Charles Scribner's Sons.

WOLIN, SHELDON
1983   "From Progress to Modernization: The Conservative Turn," *democracy,* 3: 9–12.

2

The Invention of the Past
Technology, History, and Nostalgia

JOHN F. KASSON

In his study of responses to loss and change, the sociologist Peter Marris observes, "Whether the crisis of disorientation affects only an individual, or a group, or a society as a whole, it has a fundamentally similar dynamic. It provokes a conflict between contradictory impulses—to return to the past, and to forget it altogether" (151). In this essay I wish to explore the relationship between the cultural disorientation attending rapid technological change and our sense of history. The belief that technology is a progressive enterprise leading to continual human betterment has been widely shared by Americans for at least the past 150 years. What is less often explored, however, are the implications of this belief for our understanding of history and the contradictory impulses that Peter Marris notes. The reader may suspect that as a historian I am doing some special pleading, but I wish to advance a number of interrelated arguments. First, despite the labors of more professional historians than ever before, a vital sense of history has in important ways gradually atrophied in modern America. Second, this loss of historical consciousness is inextricably connected with the particular character of America's development as a technological society and a consumer culture. Third, instead of a sense of history, we have nostalgia—which is quite a different thing. Fourth, the result is a cultural impoverishment that leaves us disoriented, bewildered, uncertain how to deal with the problems that beset us, or how to face the future. Fifth and last, this situation is not inescapable, and we must actively attempt to repossess a sense of connection and engagement with our historical past as a basis to respond creatively to change.

To substantiate such an argument, I wish to draw upon what initially may appear a motley group of examples from American history—ranging from the Colt Arms Works to Walt Disney World, from Henry James to Henry Ford—that shed light on the problem of technological change and the cultural devaluation of history.

The best place to begin is with the famous declaration of Henry Ford in a 1916 interview with a reporter from the Chicago *Tribune*: "History is more or less bunk. It's tradition. We don't want tradition. We want to live in the present and the only history that is worth a tinker's dam is the history we make today" (Rae 53). The context of this statement, it is too often forgotten, was Ford's attempt to dismiss the relevance of repeated failures in the past to the prospects for European disarmament at the time of World War I; but his remarks summarize widespread American attitudes that have stimulated technological change and in turn been shaped by it. These include a strong premium on innovation, a distrust of established procedures and ways of thinking, and a belief in progress.

Americans have been taught to pride themselves on such qualities, and in many respects they should. The sense of freedom from the shackles of tradition, the encouragement of new ideas and fresh ventures, has lain at the heart of much of this nation's historic achievement. But this dismissal of history contains important liabilities as well, which will emerge in the course of this discussion. This innovative disposition with all its virtues and vices has been a critical factor in determining the character of American technological development. For rapid technological change demands more than favorable environmental and economic factors; it requires a cultural basis. By the second quarter of the nineteenth century, the United States displayed both a high capacity to generate new inventions and, even more important, a greater willingness to adopt new technologies than any other nation in the world. As one German visitor at that time observed, "The moment an American hears the word 'invention' he pricks up his ears." Inventors such as Franklin, Whitney, Fulton, and Morse were enshrined in the antebellum pantheon of national heroes. New England textile mills scrapped machinery only five or ten years old as newer, more efficient devices took their place. Technological innovations were enthusiastically embraced in the faith that they would transform people's lives for the better. As consumers, Americans accepted standardized products much more readily than their English counterparts—whether a Colt rifle, a Jerome clock, a Waltham watch, Lowell sheets, or Lynn shoes. Low cost and simplicity of design mattered more to them than the refinements and durability associated with hand-made wares. Significantly, these manufacturers of consumer goods were among the first to develop modern, mass-production techniques in the United States (Rosenberg 31–32, 43–45; Brown 112, 132; cf. Mayr and Post; Hounshell).

Such technological dynamism altered Americans' very conceptions of history. From the time of the Puritans to the early nineteenth-century political and religious leaders had measured the health of the community in terms of the people's devotion to the virtuous standards of their forefathers; yet Americans increasingly permitted themselves a smug feeling of superiority. By the end of the nineteenth-century the sense of a moral endeavor shared with earlier generations had palpably lessened. The tradition of the Puritan jeremiad persisted, to be sure, as leaders continued to lament the waning of piety, moral and civic virtue, and social commitment. Nonetheless, in the emergent technological culture of mass production and mass consumption, the past no longer appeared particularly relevant to the concerns of the present, except as a measure of both the material and moral advancement the nation had achieved.

The most vivid example I can cite here of this changed sense of history in a culture of technological innovation is Mark Twain's novel, *A Connecticut Yankee in King Arthur's Court*, published in 1889. In many respects this story is a fable of the hubris of nineteenth-century (and we might add twentieth-century) industrial capitalism, which sought to rationalize all of society according to its values and interests. The past would appear to be the one area in which technological power cannot be retroactively applied, precisely because it *is* past, over and done with. Its conflicts, compromises, and imperfections are permanent and inescapable. But those who embrace an ideology of technological progress and control, I would suggest, often seek to capture the past in a special way. Instead of recognizing alternative possibilities, complexities, limits, and ambiguities in the historical process, they view history as the glorious foreshadowing of the contemporary order. The past appears as if through the wrong end of a telescope, our own familiar surroundings in miniature. Such a view must continually suppress rival understandings of history that explore possible alternatives in the past and that question the legitimacy and inevitability of the prevailing order. And so proponents of this ideology of progress tend to reify the past, to understand it as products more than processes, and to aim to possess it in such a way that it squares unambiguously with the values of the technological present.

The hero of Mark Twain's novel, Hank Morgan, embodies this effort. He is the epitome of nineteenth-century technological man: in his words, "a Yankee of the Yankees—and practical; yes, and nearly barren of sentiment." A superintendent at Hartford's Colt Arms Works at the outset of the story, and so one intimately involved in munitions and American mass production, he is hit so hard on the head in a fight that he is knocked back in time thirteen centuries and wakes up in Arthurian England. Now to a believer in the progressive course of history, such a backward plunge might seem unendurable torture; but as a practical Yankee he exults in his unique position: "Look at the opportunities here for a man of knowledge, brains, pluck and enterprise to sail in and grow up with the country. The grandest field that ever was; and all my own; not a competitor; not a man who wasn't a baby to me in acquirements and capacities" (20, 95–96).

In the course of the story, Hank determines to modernize Arthurian England by introducing the technological, commercial, democratic civilization of his beloved nineteenth-century Hartford. He plots the overthrow of the Round Table through use of modern technological weaponry. When the English nobility raises an army in opposition, he slaughters its ranks in one of the most painful scenes in nineteenth-century American literature. Hank installs his small crew of followers in a fortress and lovingly and meticulously assembles his era's most advanced technology of death: Gatling guns, land mines, and, his *pièce de résistance*, a row of electrified fences. When the battle is over, he announces a body count of 25,000—a final application of the quantitative standards of technological "progress" and mass production—turned here to mass destruction. Hank's putative victory, however, is at bottom a defeat. He has already destroyed his prized factories in order to save them from enemy hands. After killing the English army, he finds himself trapped in his fortress; and his supporters die infected by the rotting bodies of their victims.

Symbolically, Hank has attempted to slay the corruptions of history, to kill off the feudal past through the superior technology of the present. The past enrages him, in part, because it inevitably frustrates his narcissistic passion to control people and events, his fantasies of omnipotence which are stimulated by his enormous technological power. Once his battle with history is over, however, Hank is placed under a spell by his archrival, Merlin the magician, to sleep for thirteen centuries. It is an appropriate punishment: killing history, he is consigned to nightmare and oblivion. The impulse to retaliate for the injustices of the past yields to a sentimental desire to make restitution to it. When at last after sleeping for ages, Hank dies in delirium in the nineteenth century, he succumbs longing for reunion with the past in the person of his Arthurian wife. Without a sense of history, he is consumed by a nightmare of nostalgia and guilt (cf. Kasson 202–15).

The root meaning of *nostalgia* is *homesickness*, but as we have come most often to use the word, it is homesickness for a condition in the past rather than for a place in the present (cf. Davis). In many respects nostalgia is a modern condition, a response to rapid social and technological change that seems to sweep us away from our past, to sever our sense of historical continuity. A history of nostalgia has not yet been written, but such a work would tell us much about our modern cultural condition. In the United States nostalgic expressions had already become conspicuous by the 1830s, 1840s and 1850s, from the campaign to save Mount Vernon to such popular songs as "The Old Oaken Bucket" and "My Old Kentucky Home." By the turn of the century the nostalgic syndrome was epidemic. In personal reminiscences the theme of revisiting one's childhood home, of attempting to recover a sense of organic connection with the past severed by technological change, is especially striking. Such attempts were rife with frustration. When Mark Twain journeyed to his childhood home in Hannibal, Missouri, in 1902, he could not believe how much more shrunken it was than the commodious house he remembered. "I suppose if I should come back here ten years from now," he ruefully joked, "it would be the size of a bird-house" (Kaplan 365). The novelist Henry James suffered a still greater shock. When he returned to the United States for a visit in 1904 after a twenty years' absence in Europe, he complained he felt "amputated of half my history" by the incursions of technological society. He had become a stranger in his native land, his New York boyhood home torn down, the old sights of the city obscured or completely obliterated by the new industrial landscape, the tempo increased, the commercialism rampant. Traveling to Washington Irving's old house, Sunnyside, in Tarrytown, New York, James sensed that the very railroad which carried him hopelessly compromised his literary pilgrimage. The train symbolized "the quickened pace, the heightened fever" that cut the modern world irretrievably from Irving's gracefully indolent time and condemned the tantalized James to grasp for "the last faint echo of a felicity forever gone" (91, 156–57).

Yet such feelings were so overtly acknowledged only by a small minority. Most Americans were still encouraged to look forward rather than back, to anticipate the glories of a new century rather than to

JOHN F. KASSON

bemoan the passing of the old. So firmly had the nation embraced recent technological innovations, so completely had the goods of the emerging consumer society knit themselves into the cultural system, that many middle-class Americans could hardly contemplate life without them. An advanced technological standard of living was becoming perhaps the major gauge to measure the quality of life.

Writing in 1900, a historian of invention named Edward Byrn attempted to dramatize the extraordinary progress of the last century by demanding that his readers contemplate—even for a moment—living without it. Imagine, he said, taking a journey back in time from the beginning of the twentieth century to the beginning of the nineteenth. Leaving the present in "a luxurious palace car behind a magnificent locomotive, gliding on steel rails, at sixty miles an hour," the traveler would near the year 1800 in "a rickety, rumbling, dusty stagecoach." Gazing out the window as one traveled through the years, one would behold the milestones of technological progress passing in ghostly panorama: not only telephone, phonograph, camera, and electric lights, but as the journey lengthened, an entire catalog of modern achievement. Breathlessly, Byrn listed a long procession of inventions lost, then suddenly broke off, crying, "but enough!" as if waking from a horrible nightmare. Even to imagine truly regressing into what he called the "appalling void" of the past, there to be mired in ignorance and misery, was simply unendurable (4–6).

Byrn exemplified a new and increasingly dominant popular conception of history as a steady ascent in which technological development and human betterment marched hand in hand. Instead of finding history important as a record of human experiences, adjustments, struggles, limitations, and achievements, in which no ultimate solutions were possible, this new view presented history almost solely as the accumulation of things, and particularly the progressive development of consumer goods and inventions. It was a view of history especially congenial to the economy and culture of consumption that gradually developed in the nineteenth century and grew with astonishing speed in the early twentieth. Mass production stimulated new techniques of mass consumption, bringing myriad products within the ordinary consumer's reach. The advertising industry expanded its mission far beyond the mere publicizing of a product to the creation of new wants. New forms of entertainment, particularly movies and, beginning in the 1920s, radio, reinforced these consumer values and aspirations (cf. Harris 189–216). No wonder, then, history came to be increasingly regarded as the accumulation of goods, collectively purchased on the gigantic lay-away plan of time.

42

Which brings us back to Henry Ford, who spoke for such a view when he said, "History is bunk." Ford, as much as anyone, helped extend the consumer culture that encouraged this view of history. His famous Model T, introduced in 1908, was a car for the multitude, a machine not just for business but for pleasure. The Model T was designed for ease of manufacture, in the confidence that the buying public would readily accept a durable and economical product, without the frills of more luxurious models. As Ford reputedly said, the car was available in "any color so long as it's black."

At first the price was not spectacularly low: $850 and up. But Ford determined to bring the price down while at the same time increasing his profits, by rationalizing every stage of production, so as to make cars as much alike as "pins or matches." By developing the moving assembly line in the early 1910s, specializing machines and jobs, and driving his workers hard, Ford and his associates achieved phenomenal increases in production. By 1914 Ford workers could completely assemble an automobile in 93 minutes, compared to the more than twelve hours it had taken just a year before. By 1925 Ford had so perfected the system that it could produce a new car every ten seconds of the working day (Sward 24–43).

Such extraordinary production demanded equally extraordinary consumption, and Ford and other automotive dealers achieved it. Ford ultimately sold over 15 million Model Ts before ceasing their production in 1927. By 1929, 4,800,000 automobiles were produced in a single year, and Americans were driving more than 26 million cars and trucks: one to every five persons.

The car, Ford believed, would help carry Americans and the world at large toward a new utopia of human plenty, peace, and brotherhood by breaking down barriers of isolation and provinciality and by easing labor. "Machinery," he exclaimed, was "the new Messiah." Yet the Messiah did not come exactly as Ford anticipated. And although he never formally recanted his faith, Ford grew uneasy in the modern world he had done so much to shape. Against the swelling urban congestion and suburban sprawl which the car had helped to stimulate, Ford called for a revival of farm life. Against the image of "Flaming Youth" whirling off in chariots of freedom, he reaffirmed the virtues of the family hearth. To counter the excesses of the Jazz Age, he attempted to revive square dancing. Against the sweeping changes of the present, he immersed himself in a static conception of the past.

In 1919 Ford began to restore his boyhood home near Dearborn, Michigan. He tried to make it exactly as it had been in 1876, the year his mother died when he was thirteen. He remembered and matched the exact model of the stove in the sitting room. He remembered and matched the exact pattern and color of the parlor carpet. He dug up broken bits of china around the yard to restore his mother's dishes. He bought slippers like those she used to wear and placed them carefully under her bed. Restoration became a way of recreating his early family environment and the parental relationships he felt he had lost.

Ford's projects quickly mushroomed far beyond the restoration of his boyhood home. He soon amassed the most extensive collection of Americana then in existence. The man who had earlier snapped, "History is bunk," and had been lampooned as an ignoramus,[1] protested that he meant history *as it is taught in the school books* was bunk. What mattered were not wars and politics, so much as the common daily life, tools, and furnishings of Americans (Upwand 2; italics added). Intuitively, he rejected any conception of history that emphasized its controlling power over the present, including notions of history as a necessary burden to be carried, as the record solely of great men, or as a repetitive cycle that foreclosed new possibilities (cf. Donoghue 109–12). Ford instead sought to defend in his reconstruction a conception of history as an inventory of improvements, in which each generation built upon the works of those before. He aimed to constitute the past as a physical setting and to insist that its essence could be reassembled in the present. His ambition to control the American past, to possess it, rivalled Hank Morgan's in Arthurian England: "We want to have something of everything" (Upwand 3).

By 1929 he had prepared the Edison Institute, which included both a village of historic reconstructions and a formidable museum of technology and antiques. Intended as a tribute to Ford's great hero, Thomas Edison, the complex was dedicated on the fiftieth anniversary of Edison's invention of the incandescent lamp. The museum's exterior was a preposterous historical stage set, conflating what were touted to be "exact reproductions" of Philadelphia's Independence Hall in the center and portions of Congress Hall and the Old City Hall on the ends. In the collection of historic buildings, Greenfield Village, Ford amassed an idealized reconstruction not only of his own childhood, but of what might be regarded as the childhood of modern America. Here he placed a number of landmarks of his youth, including his father's barn, the country school he had attended as a boy, the house of his favorite schoolteacher, and ultimately his family's farmhouse. In a

1. After Ford's interviews with the Chicago *Tribune* in which he made his famous declaration, the newspaper, which supported a policy of military preparedness directly at odds with Ford's pacifism, called him "an anarchist" and "an ignorant idealist." Ford sued for one million dollars and after a farcical trial ultimately won six cents in damages.

number of other exhibits, he pursued his quest to recover the lost world of nineteenth-century youth, the shapers of that youth, and the architects of the future. He restored and installed the childhood home and bicycle shop of the Wright brothers; a schoolroom built of logs from the barn of William McGuffey, author of those schoolroom staples especially revered by Ford, the *McGuffey's Readers*; the home of Noah Webster, author of the great *Dictionary*; a summer home of Thomas Edison's childhood in addition to Edison's entire Menlo Park compound; Luther Burbank's birthplace, and so on. Occasionally, the acquisitive zeal of Ford and his collectors exceeded their historical grasp, as when in search of Stephen Foster's birthplace, they bought and reconstructed the wrong house. However much subsequent evidence came to light revealing their mistake, Ford shut his ears. It is as if the obsessive intensity of his historical effort could not slacken to deal with skeptics. He was determined to recreate in his own terms the historical and personal era he had helped to displace.

Ford's psychology is far too complex to discuss here; but his historical restoration may be viewed as an attempt at restitution—most directly to his parents, about whom he concealed deep feelings of loss and abandonment, and by extension, to the historical world of his parents, to the past. In Ford's involvement with history, as with the dying Hank Morgan of Mark Twain's novel, one can see strong themes of guilt and expiation. For both Ford and Morgan the attempts at restitution point to the sense of loss and disappointment they were trying to overcome (Jardim 158–97; cf. Sward 259–75).

What is striking, of course, in Ford's involvement with history is its emphasis upon *things* instead of more direct concern with human issues. Instead of representing a solution to the problem of keeping alive a sense of history in a technological culture, Ford's conception of reconstruction potentially made the problem more difficult. In effect he invited visitors to conceive of history "not as a series of problems and actions, but rather as a gallery of dated artifacts to be nostalgically possessed."[2] As in so much of modern mass culture, history is registered as a succession of *styles*. Surveying the past becomes an act of imaginative consumption, like shopping in a department store. The inevitable souvenir shops that accompany Greenfield Village and similar ventures satisfy on a small scale the desire to convert the past into a commodity. While searching for an alternative to the modern consumer culture, Ford's enshrinement of artifacts represented a ca-

2. Here I borrow the language of Ann Douglas, who discusses James T. Farrell's criticism of the devaluation of history (490). In my discussion of the Henry Ford Museum and Greenfield Village, I mean for my analysis to apply to Ford's original conception and not their present administration, which is much more sophisticated.

pitulation to the categories of that culture and a testament to its pervasive authority.

This version of history has proved so popular in our own time because it allows people both to have their cake and eat it: to celebrate technological progress while sentimentalizing the past. Expressions of this phenomenon are numerous, but it achieves perhaps its clearest statement in California's Disneyland and Florida's Walt Disney World. Like Ford, Walt Disney was a key figure in the creation of our modern consumer culture and one of the leading innovators in the application of technology to leisure-time activities. As early as 1940 Disney visited Greenfield Village, and the two great amusement parks he later developed significantly resemble Ford's creation. California's Disneyland, launched in 1955, and Florida's Walt Disney World, which opened in 1971, to a much greater extent than in Ford's conception of Greenfield Village, embraced history with a consumer's hug that choked the life out of it. On a scale that would have amazed even Henry Ford, Disney attempted to harmonize technological innovation with historical nostalgia. The popularity of his achievement is unquestionable. The millions of customers who throng to each of the parks yearly testify to the enormous attraction of Disney's synthesis. The nature of his success is more dubious. Without losing sight of the fact that Disney's parks are designed for pleasure, I wish to stress that the means by which this pleasure is achieved tell us a good deal about how we view history in a consumer culture.

There is no present at Disneyland or Disney World, and no sense of how past and future are connected. The five major theme areas of the parks—Main Street, Fantasyland, Frontierland, Adventureland, and Tomorrowland—are each self-contained and carefully planned to give an aura of safety. The executive vice-president and chief operating officer of WED Productions, John Hench, blandly summarized Disney's view when he declared, "Nobody worries about the past, and in a sense nobody worries about the future, because that's going to be up in space, in the space colonies. It's *today* where you have the problem," problems Disney proposed to solve through such technological marvels as EPCOT, his Experimental Prototype Community of Tomorrow (Haas 19). Thus the parks embody a historical vision calculated to reinforce the most uncritical enthusiasm for technological progess. They deny any sense of shock in historical change or even any sense of ambiguity. The theme areas assert that there are no victims in history, only winners. Not only are the most obvious victims overlooked—there is, for instance, no Wounded Knee exhibit in Frontierland—but, more fundamentally, Disney's technological creativity reassures us that in the passage of time and change nothing is truly lost,

22222222222

since what he would have us remember of the past can be magically recreated by use of space-age technology. At the same time, he promises that everything is to be gained in the future, as developing technologies bring further wonders and happiness.

This is a consumer's view of history for a corporate capitalist society, and visitors are prepared for it from the moment they enter the park. In both Disneyland and Disney World patrons arrive from the parking lots at a Gothic revival railroad station. They then proceed down a turn-of-the-century Main Street, the entranceway and preparation for all the other attractions. Every element has been meticulously designed so that visitors experience no jarring sense of contradiction. Even what Disney designers regarded as the historic impurities of architectural styles have been carefully removed. At the same time the Victorian trappings have not been chosen casually. Disney and his colleagues conceived of them as an architecture of reassurance, fusing nostalgic associations for a bygone time with the optimism of a golden age of capitalist expansion, when, in the words of John Hench, "we thought progress was great and we all knew where we were going" (Haas 18).

In this evocation of the past, the pressures of everyday life give way to a reified fantasy in which the historical and the imaginary conjoin. Along the street visitors encounter picturesque buses and horse-drawn carts, Keystone Kop musicians, inviting pastel-colored shops, Mickey Mouse and other Disney characters, and their stroll culminates at Cindrella's Castle. Like Ford's museum, Disney's Main Street is a historical stage set, but Disney and his associates did not pretend to be concerned with historical accuracy, since they believed they could surpass it in effects. Main Street was not intended as an imitation of any existing small town street; "it's what a Main Street should be." John Hench has captured the ambition of the Disney organization in insisting, "Ours is a kind of Universally true Main Street—it's better than the real Main Streets of the turn of the century ever could be" (Goldberger 40, 95).

On Disney's Main Street history is devoid of substance. It has been reduced to style, trivialized to serve as a stimulus to nostalgia. But nostalgia itself has been trivialized as well. Rather than prompting us to reflect upon attachments in our personal past, Disney uses pseudo-historical associations to intensify our attachments to his products. Public history and personal nostalgia alike have been reduced to stimulants to consumption. As one major real estate developer has approvingly noted, "What is Main Street? It is an ordinary shopping center. . . . Its purpose is exactly the same as Korvette's in the Bronx, but it manages to make shopping wonderful and pleasant at the same time. I'm sure people buy more when they're happy" (Finch 432; Goldberger 94).

As in Ford's historical reconstructions, Disney's Main Street has roots in its creator's own personal history. The son of a harsh, domineering father, Disney had a difficult childhood and, like his three older brothers, left home as soon as he could. His Main Street is an idealized version of the Midwestern towns of his youth, a tangible wish-fulfillment of the sunny childhood he never had (cf. Schickel 45–64; Thomas 26–41). In touching upon the individual psychologies of Ford and Disney, I do not mean to ascribe their idealization and trivialization of history merely to idiosyncratic personalities. My point is rather that the nature of modern culture in an age of rapid technological change is such that the universal crisis of childhood loss is compounded by a sense of historical loss. The inevitable disappointments involved in the surrender of infantile fantasies are frequently tinged by the deprivation of the material, scenes, and historical circumstances of childhood. Individual psychologies are therefore to some extent epitomes of a larger cultural condition in which we are frustrated by a lack of a sense of meaningful historical connection. The extraordinary public response to Disneyland and Disney World suggests how eagerly people turn to the ostensible solution of Disney's "Magic Kingdom."

In fact, however, Disney, like Ford and like much of modern mass culture, instead of offering a genuine sense of reintegration with history in a way that equips us to face the future, only compounds the problem of our consumer-oriented present-mindedness. The dilemma, embodied in Main Street, is most starkly evident when the visitor arrives at Liberty Square and the Hall of the Presidents.

The Hall of the Presidents was the culmination of Disney's long fascination with mechanical figures. Decades before the parks were built, he began collecting mechanical toys. These suggested to him the possibility of devising a three-dimensional equivalent to the art of screen animation. In the mid-1940s he experimented with mechanical puppets, assisted by what in Disney parlance are known as "imagineers." However, the existent technology proved too cumbersome and imprecise, and Disney shelved the idea—though he never forgot it. The dramatic advances in electronics in the 1950s and early 1960s offered new possibilities for the construction of a robot, which Disney seized eagerly. He had his workers prepare a figure of Lincoln for the Illinois Pavilion of the 1964 World's Fair, as part of an exhibition, "Great Moments with Mr. Lincoln." Disney's Lincoln was the product of complex pneumatic and hydraulic power systems controlled by magnetic tapes, technologies developed in part for the American space program in the early 1960s. But equipping Lincoln for the space age didn't prove easy. At first the "imagineers" had control problems: the Lincoln figure smashed his chair and threw mechanical fits that threatened the safety of the men working on him. Only after considerable effort was his power harnessed and his disposition made tractable (Finch 400–405).

All chief executives of the United States are represented in Disney's Hall of Presidents and all are capable of limited movements. But the Lincoln figure remains central and dominant. He is programmed for sixty-five different body movements, including seventeen facial expressions (more than those of any of the other presidents), and only he is endowed with the gift of electronic speech (Bierman 234). After the assembled presidents have been introduced and each has nodded to the audience, the Lincoln robot rises to deliver a solemn address. It is not, however, a speech that Lincoln ever gave. None was apparently deemed sufficient for such an occasion, so instead the Disney staff has pieced together various Lincoln remarks to give the audience a prepackaged "best of Lincoln." While the Lincoln figure holds forth earnestly, the other robots listen attentively and nod their assent. At the conclusion of the program, a recorded chorus of "The Battle Hymn of the Republic" swells up from behind the curtain. "Almost invariably . . . audiences break into strong applause." Then doors open along the end of each row of seats, and the audience exits smoothly and directly—to a souvenir shop! Throughout the program, the tone has been uncritically hagiographic. There is no sense that these various presidents might disagree with each other, that the past does not speak with one voice, or that we in the present might engage ourselves sufficiently with these figures to examine our differences with them. Lincoln's words have been divested of meaning, reduced to rhetorical effects. History as a whole is valued for the emotional tingle and the lump in the throat it can produce. And this emotional tension, in turn, is to be released by an act of consumption, buying a Lyndon Johnson creamer, say, or a Kennedy bust (Lelyveld). Though ostensibly an engagement with historical figures, the Hall of Presidents is really a narcissistic Hall of Mirrors of our own culture. The wonder of the program derives less from the historical careers of the various presidents than from the technological duplication of their personages. The power of Lincoln's speech depends not so much on *what* he said as he ability of a robot to mimic his words. In the guise of historical reverence and scrupulous exactitude, the exhibit divests history of any values alternative to our own. Disney's Lincoln extends the dream of mastering history through technology of Mark Twain's Hank Morgan and Henry Ford's Greenfield Village. It is ultimately a form of self-congratulation for the dominant culture of the present[3]

I have dwelt upon Disneyland and Disney World to this extent because I think that they and the other instances I have discussed are symptomatic of the cultural dilemma that makes thinking about technology and history so difficult. Beneath a superficial optimism which has long since worn thin, contemporary Americans contemplate change with apprehension because they feel historically isolated. They have imbibed so thoroughly the vision of history as a progressive drama of material accumulation and technological achievement that they look back to the past with profound ambivalence. They nostalgically ache to return to an idealized past before the onset of loss and change, but at the same time they wish to hurry toward a future where such losses no longer matter (cf. Marris). They sense intuitively that in history lie their buried roots, but they feel cut off by the terms by which progress is measured in modern life. Even the relatively recent past appears to many somewhat like Arthurian England to Hank Morgan: a distant country, poor and meager for the most part, inhabited by a few Founding Fathers and a great many cultural primitives.

Because so many Americans lack a genuine sense of alternative ways of life in the past, they cannot creatively imagine different ways of life in the future—only extensions of the technological consumer culture of the present. A striking number of people still place their hopes almost totally in technological innovations to rescue us from the problems that beset us: poverty, pollution, job dissatisfaction, overpopulation, economic development, world peace—without sufficiently recognizing these problems are far more deep-seated. They require not only fresh technological applications, but fresh political and cultural approaches as well. And while history certainly offers no easy "lessons," it can provide a complex fund of experience upon which to draw and to reflect. It is far too precious a resource to be squandered in nostalgic simplification. Instead of seeking idealized retreats, we need to recover a meaningful sense of historical connections in order to allow ourselves to live more fully and creatively in the present and to provide the basis for a sense of connection with posterity. Such a sense of connections can healthfully enlarge our awareness both of human possibility and limits. It can remind us that technological change takes place in a larger context of historical devel-

3. Two articles by Mike Wallace that came to my attention only after completing this essay treat Greenfield Village and Disney World along lines congenial to my argument, (1981, 63–96; 1985, 33–57). Pertinent to my entire inquiry but published too late for me to use is the work of David Lowenthal.

opment. It can help us look beyond the historical narrowness of the present moment and the provinciality of our consumer culture, and give us a vantage point from which to assume greater control over our lives.

Works Cited

BIERMAN, JAMES H.
1977   "The Walt Disney Robot Dramas," *Yale Review* 66: 223–36.
BROWN, RICHARD D.
1976   *Modernization: The Transformation of American Life, 1600–1865.* New York: Hill and Wang.
BYRN, EDWARD
1970   *The Progress of Invention in the Nineteenth Century.* Reprint edition. New York: Russell and Russell.
DAVIS, FRED
1979   *Yearning for Yesterday: A Sociology of Nostalgia.* New York: Free Press.
DONOGHUE, DENIS
1985–86   "Attitudes toward History: A Preface to *The Sense of Past*," *Salmagundi* 68–69: 107–24.
DOUGLAS, ANN
1977   "*Studs Lonigan* and the Failure of History in Mass Society: A Study of Claustrophobia," *American Quarterly* 29: 487–505.
FINCH, CHRISTOPHER, ED.
1973   *The Art of Walt Disney.* New York: Abrams.
GOLDBERGER, PAUL
1972   "Mickey Mouse Teaches the Architects," *New York Times Magazine* (Oct. 22): 40–41, 92–99.
HAAS, CHARLIE
1978   "Disneyland Is Good for You," *New West* (Dec. 4): 13–19.
HARRIS, NEIL
1981   "The Drama of Consumer Desire," in *Yankee Enterprise: The Rise of the American System of Manufactures,* pp. 189–216. Edited by Otto Mayr and Robert C. Post. Washington, DC: Smithsonian Institution Press.
HOUNSHELL, DAVID A.
1984   *From the American System to Mass Production, 1800–1932: The Development of Manufacturing Technology in the United States.* Baltimore: Johns Hopkins University Press.
JAMES, HENRY
1968   *The American Scene.* Edited by Leol Edel. Bloomington: Indiana University Press.
JARDIM, ANNE
1970   *The First Henry Ford: A Study in Personality and Business Leadership.* Cambridge: MIT Press.
KAPLAN, JUSTIN
1966   *Mr. Clemens and Mark Twain.* New York: Simon and Schuster.

John F. Kasson

KASSON, JOHN F.
1976   *Civilizing the Machine: Technology and Republican Values in America, 1776–1900*. New York: Viking Press, Grossman Publishers.

LELYVELD, JOSEPH
1976   "Disney's Hall of Presidents Not '76 Politics," *New York Times* (March 16): 25.

LOWENTHAL, DAVID
1985   *The Past is a Foreign Country*. Cambridge: Cambridge University Press.

MARRIS, PETER
1974   *Loss and Change*. New York: Pantheon.

MAYR, OTTO, and ROBERT C. POST, EDS.
1981   *Yankee Enterprise: The Rise of the American System of Manufactures*. Washington, D.C.: Smithsonian Institution Press.

RAE, JOHN B., ED.
1969   *Henry Ford*. Englewood Cliffs, N.J.: Prentice Hall.

ROSENBERG, NATHAN
1972   *Technology and American Economic Growth*. New York: Harper and Row.

SCHICKEL, RICHARD
1968   *The Disney Version: The Life, Times, Art and Commerce of Walt Disney*. New York: Simon and Schuster.

SWARD, KEITH
1968   *The Legend of Henry Ford*. New York: Atheneum.

THOMAS, BOB
1976   *Walt Disney: An American Orginal*. New York: Simon and Schuster.

TWAIN, MARK
1889   *A Connecticut Yankee in King Arthur's Court*. New York: Charles L. Webster.

UPWAND, GEOFFREY C.
1979   *A Home for Our Heritage: The Building and Growth of Greenfield Village and Henry Ford Museum, 1929–1976*. Dearborn, Mich.: The Henry Ford Museum Press.

WALLACE, MIKE
1981   "Visiting the Past: History Museums in the United States," *Radical History Review* 25: 63–96.
1985   "Mickey Mouse History: Portraying the Past at Disney World," *Radical History Review* 32: 33–57.

3

Artificial Intelligence
Its Accomplishments
and Prospects

MARTIN G. KALIN

Why Thinking Machines?

In the *Politics*, Aristotle speculates about using machines to replace human slaves (1130–37). The machines would try to satisfy the needs of their human masters, having figured out on their own what these needs were. Aristotle does not go into explicit detail, but implies some. To discern human needs, a machine would have to apply powers of perception and analytical reasoning to a wealth of common-sensical and maybe specialized knowledge. For example, a mechanical slave would have to perceive how profusely its master was sweating from work or sport; realize that sweat from exertion typically induces thirst; know that water and other liquids can slake thirst; recall comparable situations in an effort to infer the master's preference for this or that drink; plan how to bring the master and drink together; decide on an appropriate quantity; select a proper container to hold the liquid; and so on. Other requirements abound. The machine would need to be able to initiate movement; recognize and recover from errors of perception and reasoning; understand and generate natural language; learn; integrate various capabilities and various sources of knowledge. The list could go on indefinitely.

Aristotle's speculation might be written off to misgivings about slavery in Greek society, or perhaps to the Greek genius of anticipating discoveries far in advance. My intent is not to pick at Aristotle's motives, or to suggest that he foresaw artificial intelligence—the attempt to engineer machines capable of human-like behavior—in much the way that, say, Aristarchus foresaw Copernican astronomy. Aristotle does not talk in detail about thinking machines, but what little he says is positive in tone and rich in implication. For him, thinking machines are a way to serve human needs and desires while sparing human labor.

A Technical View of Thinking Machines

To the critic of things technological, Aristotle's account of thinking machines may seem naive, self-deceptive, or worse. Who really gains from mechanical robots—the human slaves who now become expendable, or their masters who acquire more efficient, reliable, loyal, and docile replacements? Are not thinking machines just another way in which the privileged few enforce and extend their dominion over the many? If we humans build machines that match us in intelligence or other crucial abilities, do we not thereby undermine our own position in nature? Could we wind up building machines that become our masters rather than our slaves? Even if machines can mimic us in this or that behavior, should they be allowed to do such things as care for children, dispense moral advice, pilot airplanes, or whatever else strikes us as properly and exclusively human? These cocktail-party questions are fun to ponder and abstract enough to resist compelling answers; but they are likewise frivolous when cut off from technical considerations. Before wringing our hands over the imagined economic, social, political, psychological and ethical consequences of thinking machines, we need to consider two things: (1) the specific human behavior that a machine is to replicate; (2) the technical feasibility of a machine's replicating this behavior. My paper addresses the second issue, but in general terms. It surveys artificial intelligence (AI), or the technology of "intelligent" machines; its aim is to clarify what AI now can accomplish, which is the only reliable indicator of AI's prospects. Although discussion of AI should not end at the technical level, it must start there to be intelligent.

The Diversity of AI

AI contains diverse research pursuits and applications. Some efforts in AI are commercially inspired, whereas others have a traditionally academic motivation. AI covers work on game playing, the representation and retrieval of information, machine learning, natural language processing, automated theorem-proving, machine vision, robotics, automated code-generation, planning, man-machine interface, advanced computer architecture, commonsensical reasoning, search, expert systems, and more. AI practitioners differ in orientation. Some attend to theoretical and foundational issues, addressing concerns that AI shares with other disciplines such as logic, linguistics, mathematics, psychology, electrical engineering, medicine, and economics. Others focus on applications, dealing with challenges typical of software engineering: specifying, designing, implementing, testing, debugging, delivering, and maintaining large software systems. Whether all these concerns belong under a single label is not clear, but custom keeps them there. Finally, AI practitioners come from varied backgrounds. Few hold degrees in AI as such which has not emerged as a discipline separate from computer science, itself a new kid on the academic block. Among the big names in AI are linguists, physicians, economists, philosophers, mathematicians, psychologists, physicists, accountants, and electrical engineers. AI is no monolith.

The Fuss over AI

John Haugeland imagines a press release from Florida that announces new wine made from swamp water and coal tar. The wine is reputed to be outstanding and competitively priced. Haugeland anticipates reactions that spring from predictable attitudes:

1. enthusiasm for yet another wonder of science;
2. abomination over the loss of jobs and discriminating palates, all for a product sure to be carcinogenic;
3. the debunking of a sham, because—by definition—wine must be made from grapes, fermented, and so on;
4. skepticism that coal tar and swamp water can yield anything tastier than kerosene, for decent wine requires organic molecules not found in such ingredients.

Haugeland sees, in the public debate over AI, analogs for such attitudes; and he, too, cautions that the attitudes may be premature because AI has not yet delivered its promised concoction—a truly "intelligent" computer-based system. AI has produced systems that tackle harder problems than do traditional computer systems, most of which manipulate only numerical and textual data in doing things such as analyzing statistics, generating mailing labels, tracking inventory, sorting invoices, issuing paychecks, and the like. By contrast, AI expert systems appear capable of manipulating richer and more general sorts of information, the sorts that belong to human expertise.

Such systems not only master the knowledge and experience that goes into human expertise, but also dispense and explain advice in the style of a human expert. Expert systems now do financial and medical diagnosis and recommend appropriate treatments; they configure large, complicated systems; they schedule a complex of varied tasks; they play chess at a world-class level. Are expert systems intelligent? The answer hangs on the definition of *intelligence*, of course. Instead of stipulating a definition, and then defending it, let me characterize the current generation of expert systems.

Expert Systems and Intelligence

The field of expert systems is the main AI technology making its way out of the academic laboratory and into the real world. Some have flourished there, and for obvious reasons. Expert systems, unlike human experts, can be duplicated easily and dispatched permanently to the most unappealing site. For example, one oil company has an expert system that helps diagnose and fix breakdowns in drill bits; the systems are accessible from all rigs, even remote ones in the ocean. Compared to humans, expert systems may be more reliable and consistent in performance, easier to train, and cheaper. They are not moody, late for work, or likely to unionize; and despite the popular image left by HAL from the film *2001*, even the smartest expert systems are tame.

Like a human expert, an expert system typically knows a lot about a little; unlike a human expert, it generally does not recognize clearly where its expertise stops. For example, an expert system in bacterial infections may work its way from subtle, diverse symptoms to the proper diagnosis and treatment of, say, strep throat; but the same system, confronted with clear symptoms of a heart attack, will miss the diagnosis altogether. An expert system may behave like an idiot savant; in the jargon, it does not degrade gracefully. Expert systems still amount to an impressive technology on practical and, perhaps, theoretical grounds. On practical grounds, expert systems represent the extension of computer technology into new domains; they have automated tasks that were too challenging for conventional systems. On theoretical grounds, such systems may furnish empirical support for a hypothesis that spells out the conditions for Aristotle's thinking machines. The hypothesis, from Allen Newell and Herbert A. Simon goes as follows: "A physical symbol system has the necessary and sufficient means for general intelligent action" (119). Some terms need clarification. By a *physical* system, Newell and Simon mean one that obeys the laws of physics; a system built from physical components, organic or inorganic, would be physical in this sense. A *symbol* is a physical pattern (e.g., a letter) that can be a component of a larger structure (e.g., a word). A symbol structure has a *designation* in that it denotes one or more objects: for example, the words "cat" and "mat" designate two objects, respectively. A symbol structure also has an *interpretation* which associates a given structure with a process that the system can execute. This implies that a symbol structure may behave as a command that invokes some action.

This clarification of the Physical Symbol System Hypothesis is informal, but captures its gist—that intelligent systems are achievable in any medium that supports symbols, structures made out of symbols, and some general processes that manipulate the structures. In particular, the Physical Symbol System Hypothesis implies that a computer system of sufficient complexity is capable of intelligent action; or, intelligence need not be confined to the biological world. The hypothesis is also empirical, and so challenges AI technology to deliver nonbiological systems that, through the appropriate storage and manipulation of symbols, produce intelligent behavior. Perhaps expert systems take one step, quite small, toward confirming the Physical Symbol System Hypothesis. The issue is unclear. Even if we credit one or another expert system as being intelligent inside its restricted domain, we cannot ascribe *general* intelligence to it. Newell and Simon give us a hypothesis about intelligence of a kind that Aristotle envisioned for a thinking machine, namely, an intelligence that captures the human variety in its full range.

However impressive, expert systems do *not* settle the issue of whether thinking machines are technically possible; they do not meet the challenge of the Newell-Simon hypothesis. Their intended use in the workplace underscores the point. The overwhelming majority are designed as *advisors* to humans, not replacements for them. It is clear that expert systems have extended computer technology, and equally clear that the extension falls short of the general intelligent action in the Newell-Simon hypothesis.

AI's Foundations: Search and Knowledge Representation

One way to assess AI's prospects is to look at its accomplishments in the form of expert or other AI systems. Another is to look at the engineering principles and techniques that go into such systems. This section takes the second look.

AI is a broad and diverse field, and thus hard to define. Yet two features are common to most AI systems: (1) they require explicit representation for enormous amounts of information; (2) they rely on extensive search in the processing of such information. Knowledge representation and search are foundational topics that cut across AI specializations. A survey of these topics would take us too far afield, so an example will have to do. The example is meant to give a sense of the prosaic underpinnings of an AI system.

Imagine a warehouse staffed by intelligent machines. A human supervisor would tell a machine, in ordinary language, what it was supposed to do; the machine then would fashion its own plan and execute it by drawing upon powers of perception, reasoning, and imagination as well as upon its reserves of commonsensical and specialized knowledge. Consider one small part of this scenario, namely, the machine's ability to make simple deductions. For example, suppose that the machine sees three crates stacked as follows:

> Crate A is green.
> Crate B's color is obscured.
> Crate C is blue.

The human supervisor wants to know whether a green crate sits on a nongreen one. For a human, at least one with patience for such riddles, the answer comes quickly and straightforwardly: if crate B is green, then the green B sits on the nongreen, i.e., blue C; if B is nongreen, then the green A sits on the nongreen B. A typical automated theorem-prover (Lusk and Overbeek) would reason as follows.

1. Crate A is green. (given)
2. Crate C is blue. (given)
3. A sits on B. (given)
4. B sits on C. (given)
5. For any X at all, if X is green then X is not blue. (assumed)
6. For any block X and any block Y, if X sits on Y and X is green, then Y is green as well. (This is the opposite of what the theorem-prover wants to prove; the strategy is to show that this assumption is false, which means that the desired conclusion—some green crate sits on a nongreen one—must be true.)
7. If A sits on any block Y, then block Y is green. (This follows from lines 6 and 1, by substituting A for X in line 6.)
8. If B is green then C is green. (This follows from lines 6 and 4, by substituting B for X and C for Y in line 7.)
9. B is green. (This follows from lines 7 and 3, by substituting B for Y in line 7.)
10. C is green. (This follows from lines 8 and 9.)
11. C is not blue. (This follows from lines 10 and 5, by substituting C for X in line 5.)

12. End of proof—line 11 contradicts line 2, which is given as true. Because a contradiction follows if we assume line 6, then line 6 must be false and the desired conclusion is true: there is a green block sitting on a nongreen one.

Two points merit attention. First, the proof above does not include all the dead-ends that the theorem-prover explores. For example, the theorem-prover correctly reasons from lines 3 and 6 that

If A is green, then B is green.

Yet this statement does not belong to the solution; rather, it represents wasted search for a solution. Beyond the given and assumed statements, the solution consists of six more statements; but the theorem-prover generates eighteen statements in all. The generated statements that do not occur in the solution are steps on the way to one or another dead-end. Although the numbers here are small, the contrast between the eighteen statements generated, and the six used in the solution, illustrates the general problem of search.

A second point about the proof deserves mention. The solution's line of reasoning is short but tedious, and hard for us humans to follow. We reach the same conclusion by a more direct route, namely, by noting that crate B's specific color does not matter: whether B is green or not-green, a solution follows. Yet the example highlights important principles. One is that machine "intelligence" may operate quite differently from the human kind, even when the two achieve the same behavior. The other is that getting a machine to solve even a trivial problem is not itself a trivial problem, for a machine must be equipped with explicitly represented information ("declarative knowledge") and the recipe ("procedural knowledge") to fashion a solution out of this information. On a more technical level, the example illustrates how both knowledge and search are involved in AI systems. The automated theorem-prover searches through various substitutions of A, B, and C for variables X and Y until it finds a sequence of substitutions that produce a solution. The example gives one sequence of substitutions that work; sequences that failed have been omitted, but can be imagined easily. (For instance, try substituting B for X and C for Y in the first step; then try C for X and C for Y; and so forth.) The theorem-prover needs not only the information given in the problem statement which details how the blocks are arranged; it also requires the knowledge that a green crate cannot be blue at the same time.

AI and High-Level Representation

There are higher level (namely, more human-like) ways to represent the crate problem which lead to a more efficient solution than the brute-force search depicted above. Instead of dealing only with particular colors, such as green and blue, it would be both more human-like and efficient to have the robot deal with the notion of color itself as a set that includes instances, but also decomposes into subsets, such as the set of all colors but green. This way of looking at things leads to a quicker solution if crate's B exact color is left open, and seen instead as belonging to one of two sets: the singleton set that contains only green, and the set that includes all colors but green. In this case, picking B's color from either set gives a solution. Under this approach, the automated deduction is one step long (McSkimin and Minker 231).

That AI looks for human-like representations is not surprising. It is trying to make machines duplicate aspects of human behavior; so human behavior serves at least as an implicit model for the machine replication. Further, problems such as planning or natural language processing are hard enough to solve when represented at a high level; they might become prohibitively difficult if represented in a less natural fashion. Finally, high-level representations may constrain the machine's search for a solution, and thus promote efficiency.

High-level representations, whatever their rationale, suggest kinship between organic (specifically, human) and machine intelligence. For if machines and organisms represent and process information in similar ways, then perhaps they are similar in underlying structure. AI thus reinforces the idea, introduced before AI, that the human mind can be regarded as a mechanism that executes programs—namely, as a computer. We know already that computers can be fashioned out of various materials. Using electric circuits made out of silicon is but one way to make such machines; electrochemical circuits made out of protoplasm may be another. This kind of talk evokes sharp reactions to the effect that humans and computers differ essentially, and even provokes rash predictions about the limits of what computers can do. The next section tries to sort out this controversy.

The Computer as a Cognitive Model

Computer-based systems can solve certain sets of complex mathematical equations far better than any human; give expert advice on a variety of topics; play games such as chess at a world-class level. No such systems can invent sets of complex mathematical equations that are important enough to solve, can learn expertise through experience, or invent a game as challenging as chess. Further, there are no "intelligent" machines that can juggle eggs, climb stairs, or straighten up closets as well as a human can. So on the issue of intelligence, man-machine comparisons have mixed results. The ordinary capabilities of humans are very hard to replicate, and some of the supposedly extraordinary ones are quite easy. Other capabilities, especially those associated with imagination, invention, and learning, also are very hard to duplicate.

What are implications for the study of intelligence, in both human and machine? Does AI research suggest that the "lower-level" behavior of humans, the sort associated with ordinary sensory and kinetic powers, is the most distinctively human, as this behavior is so hard to replicate? Does it likewise suggest that behavior associated with intellectual powers is the least distinctively human? Where then are the powers of imagination, invention, and learning to be put? Do they belong among the "low-level" or the "high-level," or perhaps somewhere in the middle? Do the results of AI, paltry as they are in comparison to AI's ambitions, suggest that machines are condemned to remain less intelligent than humans? Do they instead suggest that machines are destined to become more intelligent than humans? Or do the results currently suggest that machine intelligence resembles the human variety on some points, and contrasts sharply with it on others?

These questions are not rhetorical, although they emphasize again AI's status as an emerging technology that, in time, may become an established science. Once established, with solid results under its belt, AI may be in position to speak with authority to the other disciplines that deal with human beings, intelligence, and some mix of the two topics. It is far too early to tell whether AI and its allied disciplines, such as cognitive science, will have a great impact on what we humans believe about ourselves and our place in the scheme of things. Nonetheless, even infant AI has threatened some traditional beliefs and attitudes. We once believed our planet to sit at the physical center of the universe, but Copernicus and others eventually corrected this misconception. At one time we thought ourselves distinct in origin from the other animals; evolutionary science disabused us of this notion. For a time thereafter we imagined ourselves as standing apart from the other animals in alone making tools; zoological and related sciences corrected the error. It is now unclear whether only we humans use symbolic language for communication.

AI, in time, may produce machines capable of behavior that now strikes us as distinctively human. Perhaps it will produce machines fluent in natural language, or ones capable of learning from both formal instruction and ordinary experience. Maybe there will be machines that not only think in some sense, but reflect as well—know that and what they know. If machines ever prove capable of these and other activities once believed exclusively human, then AI will have given us a new view of ourselves and our place in nature. When the arrival of such machines seems imminent, the debate should begin in earnest about AI's social, political, psychological, economic, and ethical implications. Until such time, the debate should be recognized and enjoyed as recreational.

AI's eventual results, however impressive, may not be taken for what they are. Haugeland's analogy of Everglades wine is true to history, notably its emphasis on debunking as a knee-jerk reaction to any new discovery that unsettles conventional beliefs and attitudes. (Recall the papal astronomers who declined to look through Galileo's telescope, for they already *knew* the number and position of the planets.) AI may deliver machines that replicate much of what now passes as essentially human behavior only to face quibbles about whether such machines deserve to be characterized as thinking, intelligent, and the like. The issue is not crucial, at least for AI. Its primary task is to fashion machines that, like Aristotle's mechanical slaves, can duplicate some or all of human behavior. How much behavior should these machines be allowed to duplicate? If AI technology ever makes this a real issue, its debate will take us well beyond the technical matters of interest in AI proper.

Conclusion

I have encouraged caution in assessing AI's prospects, and I should practice what I preach. Yet a summary of points made above is less inviting than a bit of "crystal-balling." Let the reader be cautious instead.

AI is not mainstream computer science, but an exploratory or experimental technology with a few impressive, scattered results. This will change. AI tools and methods of problem solving, from its languages to programming techniques to specialized hardware, will be incorporated into what now passes as "conventional" computer science and technology. AI will show up in the workplace not so much as independent systems, although these will increase in number, but rather as components in systems that draw on many other technologies as well: data base, signal processing, graphics, and much more. AI will become part of the automated solution to increasingly complex problems. In time, AI will stop going under a special name because here, too, success will mean loss of identity. The overall result will be

the automation of tasks once thought too intricate or otherwise difficult for machines to handle; gradually, humans will regard the tasks
thereby automated as somehow less than human. Is hauling stones
properly human behavior? Is computing cube roots? The answer depends largely on the technology at hand. Computer technology, by
integrating components once separated out as AI, will continue to re-
define the sort of behavior that we regard as specifically human.

In earlier times, the telescope was a technology that brought trouble
to its inventor, Galileo, by threatening inherited, officially sanc-
tioned—and erroneous—beliefs; today it is a common and quite uncontroversial technology. I hope that thinking machines will become
as commonplace and uncontroversial, although the many technical
problems temper this hope. If a machine capable of general intelligence ever comes to pass, my guess is that it will come after others
with highly specialized intelligence that surpass anything done with
current AI technology. If we come to build machines with special and
even general intelligence, we will learn to accept and to exploit the
differences between them and us. Such machines then will seem less
threatening, more natural. If and when the thinking machine is as
ordinary as the telescope, the present-day critic of AI will seem a
latter-day papal astronomer.

Works Cited

ARISTOTLE
    1964   *Politics*. In *The Basic Works of Aristotle*. Ed. Richard McKeon.
                New York: Random House.
HAUGELAND, JOHN
    1985   *Artificial Intelligence: The Very Idea*. Cambridge: MIT Press.
LUSK, E.L., and R. A. OVERBEEK
    1984   *The Automated Reasoning System ITP. Argonne National Labo-
                ratory Report ANL-84-27*. Argonne, Ill.: Technical Reports.
McSKIMIN, J. R., and J. MINKER
    1979   "A Predicate Calculus Based Semantic Network for Deduc-
                tive Searching." In *Associate Networks*: 205–38. Ed. N. V.
                Findler. New York: Academic Press.
NEWELL, ALLEN, and HERBERT A. SIMON
    1976   "Computer Science as Empirical Inquiry: Symbols and
                Search." In *Communications of the Association for Computing
                Machinery* 19: 113–26.

4

Conflicting Models of Rights
The Case of
Technical Obsolescence

ROBERT ALLAN COOKE

In a rapidly changing society, technological innovation tends to be perceived as either an organ for desired change or as a threat to existing normative values. Whether a blessing or a curse, technology introduces important ethical and social issues.

One such issue is technical obsolescence. I refer here to the problem that arises when certain individuals within a firm or certain segments in the work force possess technical expertise made obsolete in the wake of technological change. This problem is not new. For years technological innovation and the related requirements of new occupational skills have led to the widespread elimination or redistribution of jobs. In other words, technical obsolescence invariably leads to either job displacement or to job retraining. Both alternatives raise the fundamental question whether firms have any moral obligation to lessen the impact of such displacement or to provide programs that upgrade technical expertise.

In this paper, I will briefly examine some basic arguments for and against such moral obligations. By constructing models based upon the work of John Rawls and Robert Nozick, I conduct what may be described as a rights analysis, an examination of whether individuals have certain entitlements that imply corresponding moral obligations for others. After applying these models to the issue of technical obsolescence, I claim that employees generally are not entitled to protection from job displacement and employers do not have any moral obligations to lessen the impact of such displacement under normal conditions. However, I do acknowledge special conditions under which an employer may have such obligations.

The Social Good Model

In an exhaustive analysis of justice, John Rawls argues that a just society is one which upholds the principle of fairness. Rawls observes that a liberal democracy, if it is just, must justify any social and economic inequality. This is done by introducing two elements to ensure fairness, the difference principle and the social minimum. Fairness dictates that such inequality is allowed if and only if this arrangement benefits the least-advantaged members of society and if and only if a socially acceptable standard for distributing social and economic goods is established and maintained. Although inequality exists in the United States, the average person still feels the country is just because its economic and social arrangements provide more opportunities for individual advancement than alternative systems. This applies to the poorest in our society as well as the richest. To ensure a basic quality of life for all citizens, a social minimum has been established through such mechanisms as social security, minimum wage, and food stamps. Just what constitutes fairness in such procedures is determined by rational, disinterested deliberations (Rawls 313).

According to this model, entitlements are important mechanisms in contributing to the overall *social good*. They are embedded in a scheme of just arrangements determined by rational deliberation. These arrangements, in turn, are bounded by the principles of liberty and equality. Social or economic inequality is acceptable as long as everyone is in some way compensated for such inequality and is protected by an established social minimum. The social minimum is essential, for it preserves a necessary level of individual dignity and self-esteem. For Rawls, this is critical since self-respect is the most important primary good (440).

Thus, one is entitled to self-respect, and a just society obligates each of us to respect such an entitlement and the social and economic arrangements that promote self-esteem. Even inequality is acceptable as long as it provides opportunities for increasing one's sense of worth, whether rich or poor. In other words, even though people possess different natural assets, each is not allowed to make unrestrained use of those assets without considering how their use will affect the social good.

We see then the difference principle represents an agreement to regard the distribution of natural talents as a common asset and to share in the benefits of this distribution whatever it turns out to be. Those who have been favored by nature, whoever they are, may gain from their good fortune only on terms that improve the situation of those who have lost out. . . . But it does not follow that we should eliminate these distinctions. There is another way to deal with them. The basic structure can be arranged so that these contingencies work for the good of the least fortunate. (Rawls 101–2)

According to this perspective, entitlements are not based upon some conception of unrestrained natural liberty that invariably allows those with favored natural assets to arbitrarily gain advantage. Rather, a just society provides a rational scheme that treats entitlements as governed by the social good. This is the moral point of view that enhances self-esteem or self-respect.

The Natural Liberty Model

By contrast, Robert Nozick develops a concept of rights that heavily relies on Locke's notion of "justice in holdings."

If the world were wholly just, the following inductive definition would exhaustively cover the subject of justice in holdings.
1. A person who acquires a holding in accordance with the principle of justice in acquisition is entitled to that holding.
2. A person who acquires a holding in accordance with the principle of justice in transfer, from someone else entitled to the holding, is entitled to the holding.
3. No one is entitled to a holding except by (repeated) applications of 1 and 2.
The complete principle of distributive justice would say simply that a distribution is just if everyone is entitled to the holdings they possess under the distribution. (Nozick 151)

In other words, what a person is entitled to is determined by a process of the just acquisitions of goods or the just transfer of goods. One legitimately acquires a previously unheld good when he mixes his labor with it and is able to maintain the holding acquired. Goods or holdings acquired through theft, fraud, coercion, and related activities are not legitimate entitlements. Those who acquire goods these ways violate the principle of justice in acquisition. Similarly, transfers of goods to others are legitimate if the person transferring said goods is entitled to them.

Notice that the individual moral agent is the key to the entitlement process. Interference in this process from structures or arrangements that determine a just distribution of goods is not acceptable. To paraphrase Nozick, individual rights are so far-reaching and strong that the only legitimate function of the state is to protect citizens from force, theft, and fraud and to enforce contracts. Any more than this violates an individual's right not to be forced to render aid to others (ix).

This is the case even if the concept of the social good is used to counter such claims. The argument that fairness dictates arrangements securing the social good is unacceptable for those subscribing to the Natural Liberty Model. For Nozick, fairness is suspect because Rawls and others claim that fairness may require an individual to sacrifice holdings in the interest of the disadvantaged. Natural liberty requires consent or choice on the part of the individual being asked to transfer goods he is entitled to. To act otherwise in the name of social good undercuts the strong notion of rights underlying this model. This is true even if that same individual has benefited from the existing social arrangements. "So the fact that we partially are 'social products' in that we benefit from current patterns and forms created by the multitudinous actions of a long string of long-forgotten people . . . , does not create in us a general floating debt which current society can collect and use as it will" (Nozick 95). Even if Rawls' principle of fairness could be reformulated to avoid this criticism, it would still be necessary for an individual to consent to constraints being placed on his holdings.

Of course, what is at stake in the Social Good Model is the concern that individuals with superior natural assets will create an unjust distribution of goods or holdings. Using one's natural assets without conforming to the needs of the social good is unacceptable because it prevents a fair distribution of goods. Such a skewed distribution is arbitrary from a moral point of view.

By contrast, those advocating the Natural Liberty Model see nothing wrong in deriving benefits from natural assets. It is in keeping with the principles of justice in acquisition and transfer. The argument is made that because holdings flow from natural assets, people are entitled to them. In other words, even if an individual's natural assets are arbitrary from a moral point of view, he is entitled to them and any holdings or goods derived from those assets (Nozick 225–26). The thrust of the Natural Liberty Model is best summarized by this dictum: "From each as they choose, to each as they are chosen" (Nozick 160).

The Application of the Models

I have spent considerable time outlining the basic elements of both models not only to present an accurate presentation of rather complex arguments but also to provide enough detail for the reader to judge how each model would address the problem of technical obsolescence. Let us examine each model.

A. The Social Good Model (SGM)

The main lesson learned from SGM is that the operational principle of fairness is needed to curb the otherwise unrestrained acquisition of goods by those arbitrarily blessed with superior natural assets. Such restraints insure the fair distribution of goods for the benefit of the least-advantaged. This is critical to the promotion and enhancement of self-esteem or self-worth for all. As a primary social good, self-worth is valued above all else. Anything which compromises self-worth poses a threat to the just society.

Does this analysis shed any light on the issue at hand? I think so. One of the most important sources of an individual's sense of worth comes from what he produces or creates. Since such satisfaction and productivity are bound up with employment, the linkage is obvious. If employment significantly contributes to self-worth, then job displacement undercuts this and is unacceptable. Similarly, if technical upgrading enhances an individual's ability to maintain employment or increases his sense of self-worth, then it is a positive good that should be promoted.

This is true whether or not there is rapid technological innovation. Does this mean that a corporation has a moral obligation to lessen the negative effects of job displacement? At the least, it does. Just what steps should be taken are open to debate. Certainly, most advocates of this position feel that some form of job retraining is required. Others claim that job placement should accompany such measures. These measures, coupled with severance packages that cover the essential costs of living for a reasonable time period and provide some health care provisions, could significantly lessen the negative effects of job displacement. Some firms may even set up special funds to handle special cases.

Yet SGM goes much further than this, for the bottom-line issue is whether an individual has a moral right to employment and protection from job displacement, whatever the cause. This is an issue that reflects the basic difference between SGM and NLM. Both models claim that liberty is a primary principle of a just society. Both models recognize that certain individuals with naturally superior skills will acquire more holdings than others. Yet in SGM this liberty must be tempered with equality and the underlying principle of fairness. The social good determines the degree to which individual freedom is exercised through the difference principle and a basic social minimum. Such constraints are justified because natural assets are morally arbitrary.

The most lucid argument for claiming that employment is a universal human right comes from James Nickel in his article "Is There a Human Right to Employment?" Nickel clearly refers to SGM, though not by name, when he describes the argument that there is a human need for employment because: (1) employment is necessary to acquire the essential goods of life; (2) employment leads to self-respect and the respect of others; and (3) employment is an important means of self-development (250–51). If true, this argument shows that it is in the interest of each individual to have a job, if he is able. But does this demonstrate that a person has a right to a job?

Nickel suggests that such interests are indeed rights because: (1) employment is so critical for survival and self-esteem that it should be treated as a human right; (2) employment is the only fair and just way to compensate the disadvantaged for the distortions created by class interests and class prejudices; and (3) job displacement results in such a severe loss of dignity and respect that it must be avoided at all costs in order to preserve the common good (252–53). Thus, any interest that is as crucial to human dignity is a candidate for being considered a moral right. This argument for claiming employment as a right is so strong, according to Nickel, that any objections pale by comparison.

Of course, such obligations may be costly. If the firm is unable to fully meet such obligations, then society has a responsibility to resolve the problem. This can be done through cooperative ventures by various business and labor organizations. Moreover, government agencies can marshall the necessary resources. CETA and the Manpower Training Act are examples of such efforts.

B. The Natural Liberty Model (NLM)

Instead of relying on a Rawlsian view of fairness, NLM stipulates that freedom of choice should be protected even if this freedom sustains economic disparities. Although natural assets may be arbitrary from a moral point of view (unless one subscribes to the concept of Divine Providence), the bottom line is that an individual is entitled to any goods or holdings that are derived from natural assets. This is not a matter for public debate. He is entitled to such benefits as long as just procedures are used in the acquisition and transfer of goods. To claim that the resulting distribution is somehow unfair is a bogus argument, for fairness emerges precisely from the implementation of just procedures.

In fact, the concept of fairness espoused by those advocating SGM appeals to a vague notion of the social good. Invariably, this is used to legitimize steps that redistribute goods to benefit the least-advantaged. For NLM such a process is inherently unjust because it undermines freedom of choice, the basic right of all individuals.

The claim that those individuals with superior natural assets will somehow acquire goods to the detriment of others is a simplistic argument. Certainly, if detriment simply means that some will have more than others, this is trivially true. On the other hand, if detriment means that society will suffer from such an arrangement, NLM rejects this claim. In point of fact, this arrangement will benefit all members of society, for such an arrangement will motivate those with natural abilities to be more productive. The resultant increase in capital and productivity will lead to more employment, a larger tax base, and more philanthropy. This in turn will benefit those unable to be as productive. Does this mean that all who are blessed with superior natural assets invariably act to benefit others? Certainly not; yet those who endorse NLM believe the vast majority of gifted individuals improve the quality of life through their creative efforts.

Moreover, the principles of justice in acquisition and justice in transfer have built-in safeguards including the following procedures: An individual must maintain control over the good(s) acquired, and an individual cannot legitimately exercise exclusive control over goods that are essential to the preservation of life. If such conditions are met, the disadvantaged have no moral claim on the benefits in question.

Of course, one might object to this analysis by claiming that it ignores the historical fact that individuals have often benefited from distortions in the distributive process. NLM recognizes this fact. Even so, inequities do not mean that any individual has some moral debt to repay those who did not benefit from such patterns. The past is past, and we cannot change it. If an individual feels there is a debt to repay and proceeds to do just that, then there is no quarrel with that decision. The only legitimate way in which a redistribution of goods can occur is through individual consent. If an individual is unwilling to do this, he cannot be forced to do so in the name of some vague notion of moral obligation. There is no floating moral debt to the disadvantaged. Does this analysis shed any light on the basic question of this paper? I think so.

According to NLM, the answer is straightforward. What needs to be recognized is the fact that employees are not the only individuals affected by technological innovation. The owners of firms also have a stake in the outcome. They have taken risks to remain competitive in a rapidly changing environment. They put their goods or holdings on the firing line every day. If conditions change and job displacement is unavoidable, they take no delight in this. They have the right to protect their investments, even if layoffs ensue. It should be noted that the owners of most large firms are not just a few wealthy executives who manage the company. The majority are stockholders with varying levels of investment, notably pension funds and mutual funds. As long as job displacement is necessary and is not capricious, there is no moral issue to debate.

Moreover, when an individual exercises choice, he bears some responsibility for the consequences of that choice. Of course, there is an element of risk involved in any choice. It is no different in employment. When an individual seeks employment, he is taking a risk. He is betting that he has the necessary qualifications to secure the job. Even the most optimistic individual recognizes an element of luck in the competitive process. If an individual is successful in securing the desired job, there is no guarantee the job will last indefinitely. Under normal circumstances, there is no lifetime guarantee of employment. The fact that technological innovation may hasten technical obsolescence and job displacement is significant, but it does not mean any moral obligation follows from this. Others may see such displacement as lamentable, but that does not entitle them to force the firm to retain those employees.

Similarly, those advocating NLM recognize that self-worth is an important aspect of employment. Any individual's sense of self-worth should be promoted and enhanced when possible. Yet, this does not mean that a firm has a moral obligation to avoid all actions which may undercut a sense of self-esteem. If a firm feels it is desirable to do something and voluntarily takes the necessary steps to avoid displacement, then that is commendable. But it is not obligatory. Employment is a two-way street. Any individual has the choice of seeking new employment that may enhance his sense of self-worth. In a free society, job mobility is important. We would not want to constrain that individual from choosing another job any more than we would want to constrain an employer from determining appropriate employment policy.

In other words, those who subscribe to NLM do not believe that any firm has a moral obligation to lessen the impact of technical obsolescence, even if this involves job displacement. Similarly, there is no moral duty to upgrade technical skills for any employee. NLM rejects any claim(s) that employment is a human right. Just because employment is a social good does not mean that it is a basic right. After all, the basic right of free choice applies to employers and employees alike.

The Question at Hand

In this paper, I have reviewed the ethical implications of technical obsolescence in light of two competing models. Now I wish to examine SGM and NLM more closely. Several observations are in order.

First, in a just society, it is important to have a vision of the common good. No rational person really denies this. Certainly, SGM is based upon such a vision. It is also true that self-respect and human dignity are hallmarks of a just society. The issue at hand is just how such dignity is to be achieved. According to SGM, it is best realized through social arrangements that redistribute goods in ways that benefit even the poorest in society.

This argument may sound plausible, but it diminishes the importance of free choice. Self-respect and dignity are impossible without freedom of choice. When consent is lost or not given, the individual loses his sense of control over the situation. This, in turn, adversely affects his sense of self-respect. If autonomy is diminished for any individual, it should be of concern to the entire community. That is why a just society puts such a high premium on freedom of choice. Even those advocating SGM recognize the importance of consent; yet their vision of the common good allows for an abridgement of consent in the name of fairness. This not only disregards the essence of human dignity, it also reveals a type of paternalism that lacks real respect for the individual. In other words, those advocating this position do not really believe that the individual is capable of making choices without a careful monitoring. Such an attitude undercuts the freedom of choice that is the hallmark of a just society. Those who would claim that such paternalism is based upon fairness do not understand the meaning of the term.

Second, freedom of choice leads to the acquisition of goods or holdings. Undoubtedly, those with superior natural assets are often in a better position to acquire goods than those individuals with inferior natural assets. Yet, as long as there are fair procedures for acquiring and transferring goods, justice is preserved. This is true even though those who have such assets may not deserve them in any moral sense. An individual should not feel guilty that he has superior assets. He should take advantage of the competitive edge it brings. After all, each individual has a different starting point in life. Some are better off than others. The measure of a person's worth is often measured in terms of what has been accomplished in spite of initial adversity. The fact that people have different starting points in life is normally not a moral issue; it is simply a correct observation. One does not have a moral obligation to correct accidents of birth. To penalize an individual for having superior natural assets by restricting his usage of those assets is blatantly unfair. People do deserve the goods or holdings that follow from their utilization of natural assets, as long as just procedures are followed. The argument that the common good justifies such restrictions is simplistic at best.

Similarly, certain individuals may benefit from social distortions of the past. These distortions may have unfairly distributed goods. Yet the fact that these distributions were initially unfair has little bearing on the present.

Third, those who advocate SGM point out that employment is one of the most important sources of self-respect and dignity. This coupled with several related arguments allegedly shows that employment is a basic human right. No rational person denies that employment is closely aligned with self-respect. It is in the interest of every individual to have meaningful outlets for expressing individual worth. A job is a significant way of doing that. This is not a trivial truth for the individual or for the community. On the other hand, it is one thing to say an individual has a strong interest in something and quite another to claim it as a right.[1] A right or entitlement is a very basic concept that should not be trivialized. NLM provides a correct analysis of such rights. In other words, no matter how important employment is to an individual's sense of worth, it does not mean that he is entitled to a job. The claim that employment should be considered a right on consequentialist grounds requires an explanation as to how this alleged entitlement overrides natural liberty or choice. In addition, he must explain why it is fair for an individual to change jobs but unfair for an employer to terminate an employee. This is an inconsistency SGM cannot resolve.

Fourth, there is the question of risk. SGM attempts to minimize risk for the least-advantaged by devising social arrangements that constrain the holdings of those blessed with superior natural assets. This is both unfair and inconsistent. It is unfair because there is no genuine concern for the interests of those with superior assets. Such individuals take risks, but the proposed arrangements do not reward them. If anything, they are penalized for their efforts.

It is also inconsistent. On the one hand, society needs individuals

1. The argument that an interest may become a right has interesting parallels with the "is-ought" controversy in ethical theory. The problems attached to the "naturalistic fallacy" may also apply to the argument we are examining.

with superior assets to take risks. This is the fundamental driving force behind progress. Such progress leads to greater capital formation, greater availability of goods and services, and greater prosperity for all. Even Rawls concedes that the poorest individual benefits from such an arrangement, provided that the social minimum is maintained.

On the other hand, by constraining individuals with superior natural assets, SGM removes the necessary incentives for taking great risks. This in turn has a negative impact on capital formation, the availability of goods, and the level of profit. In other words, the constraints developed in the name of the common good may actually work against it.

When all is said and done, the bottom line is this: firms do not have any moral obligations to lessen the negative effects of technical obsolescence. Though employment is an important part of an individual's sense of worth, the firm is not morally obligated to retain the employee when change occurs. This may seem callous, but in a free society consent is paramount. Choice always involves risk. Sometimes you win, sometimes you lose. In either case, both the employer and the employee must assume responsibility for the choices made. this holds true for both the employer and the employee.

Of course, if a firm's economic interest leads it to upgrade the technical skills of its work force or to provide lifetime guarantees of employment, it should be commended for doing something that is also socially desirable. This type of action is not a model of moral obligation; rather, it is a prudent response to what is deemed important by society in general or by shifting public opinion that may or may not conform to standard ethical guidelines. Moreover, there is disagreement as to how one determines what is socially desirable and what measures should be taken. In any case, the value of acting to meet certain perceived societal needs is enhanced by the fact that freedom of choice is exercised, not coercion.

Parting Thoughts

My line of reasoning may seem callous to many; but a just society is one in which an individual is free to fail as well as to succeed. Does this mean that no firm has ever had nor will ever have any moral obligation to lessen the negative effects of technical obsolescence? No, it does not. Under normal circumstances, no such obligation is present. Yet, an obligation may exist if:

1. an employee has continued to maintain and operate outdated equipment out of loyalty to the firm despite oppor-

tunities within the company to train on the newly acquired equipment;

2. the firm has intentionally dissuaded employees from upgrading technical expertise by assuring them that their jobs would be secure despite technological change; or
3. preexisting agreements or promises are in force.

In the first instance, an obligation may exist because of an apparent implicit promise. Out of loyalty the employee has labored on outdated equipment to smooth the transition to newly acquired technology. He does this with the expectation that his job is secure. Is this a reasonable expectation? It would seem so. He has made a choice based upon information learned in the present situation and in years of service to the company. He has done this in spite of opportunities to upgrade his technical expertise. After all, loyalty is a two-way street.

In the second instance, an obligation may exist because of an apparent explicit promise. The promise of job security affects one's view of job mobility. If an individual is content with the job he has, then he is often reluctant to look elsewhere. This is especially true if job security is no longer an issue. What remains to be determined is the extent and nature of the promises made.

In the third instance, the situation is almost identical to that of the second. The major difference is that the agreements or promises are formalized. Disputes here are settled within a legal framework.

Thus, a firm may have a moral obligation to prevent job displacement or to upgrade technical expertise if any of these three conditions apply. These are the only exceptions. Each must be settled for that specific case to determine whether a moral obligation exits. Under normal circumstances, no such moral obligations apply. This may be regrettable, but it is true.

Works Cited

NICKEL, JAMES W.
1986   "Is There a Human Right to Employment?" In *Philosophical Issues in Human Rights: Theories and Applications*. Edited by Patricia H. Werhane, A. R. Gini, and David T. Ozar. New York: Random House.

NOZICK, ROBERT
1974   *Anarchy, State, and Utopia*. New York: Basic Books.

RAWLS, JOHN
1971   *A Theory of Justice*. Cambridge: Harvard University Press.

PART TWO

Work

5

Technology and Its Critics
The Degradation of
the Practical Arts

CHRISTOPHER LASCH

Taylorism Revisited

The prevailing view of the relations between technology and values can be simply stated. Technology is ethically neutral. It is a tool that can be put to good uses or bad uses. How it is used will depend on our values, not on technology itself. Instead of talking about technology, which is simply a given, a normal aspect of our everyday environment, we would do better to talk about values and about the possibility of humanizing the industrial order. We would do better to remind ourselves, in other words, that machines, after all, are merely servants of the human beings who design them.

One of the founders of a high-technology communications company expressed these opinions in a recent interview:

> Technology is absolutely neutral and the same microprocessors [can] be used for good or evil. The determination of that really comes through what the individual, what the collective group of individuals, align themselves with. What our values are, what priorities we have in life, I think that's what the real question is, not the technology.

The same ideas appear in a full-page advertisement for the international company United Technologies, entitled "Technology's Promise": "Ethically, technology is neutral. There is nothing inherently either good or bad about it. It is simply a tool, a servant, directed and deployed by people for whatever purpose they want fulfilled."

It tells us something about this way of looking at the issue that it so often appears under corporate sponsorship. It is an interpretation designed to provide reassurance and to create the illusion that we all have a share in deciding how new technologies will be used. Note the vagueness of this talk about "individuals," "collective groups of individuals," and "people," which discourages us from asking just which people in particular design and control our technology, which people are served by it, and which people, on the other hand, stand to lose by the continuing development of this technology along its present lines.

An examination of the impact of technology on the transformation of work and the changing class structure of industrial society dispels the illusion that technology is a neutral and impersonal force. It is misleading even to speak of the impact of technology on the work process, since this formulation implies that technology originates outside the work process—in the laboratory, presumably—and has an "impact" designed or anticipated by no one in particular. In fact, much of modern industrial technology has been deliberately designed by managers for the express purpose of reducing their dependence on skilled labor. One of the early architects of this technology, Frederick Winslow Taylor, the founder of scientific management, spoke much more candidly about technology and its implications than corporate spokesmen and their propagandists speak today. In his book, *The Principles of Scientific Management*, published in 1911, he described a struggle for control of production between management and workers. The success of scientific management, as Taylor saw it, depended not so much on the introduction of machinery as on the managers' expropriation of the craft knowledge formerly controlled by the workers— "this mass of traditional knowledge," as Taylor called it, "a large part of which is not in the possession of management." That the workers understood the implications of Taylor's reforms is shown by their resistance to them. As Taylor noted, his attempt to redesign the work process, to deprive workers of any technical initiative, and to reduce them to the position of carrying out orders issued by the planning department, "immediately started a war, . . . which as time went on grew more and more bitter. . . . No one who has not had this experience can have an idea of the bitterness which is gradually developed in such a struggle." In a remarkable passage, Taylor admits that the workers' resistance to scientific management was well founded. When workmen, formerly his friends, asked him, "in a personal, friendly way, whether he would advise them, for their own best interest, to turn out more work," he "had to tell them," he says, "as a truthful man . . . that if he were in their place he would fight against turning out any more work, just as they were doing, because under the [new] system they would be allowed to earn no more wages than they had been earning, and yet they would be made to work harder" (32, 49–50, 52).

The movement for fully automated industrial production in our own time, sometimes referred to as the second Industrial Revolution, originated in a struggle for control of production in the years immediately following World War II, under conditions remarkably similar to the conditions that earlier had inspired Taylor's scientific management. "What is today called 'automation' is conceptually a logical extension of Taylor's scientific management," Peter Drucker has written (26). Taylor had attempted to separate the planning from the execution of tasks; but his innovations had achieved only partial success. At the end of World War II, many industries continued to depend heavily on skilled labor, notably the machine-tool industry itself, and the workers still controlled the pace of production. The rapid growth of industrial unionism under the New Deal, moreover, made workers increasingly resistant to managerial authority. In a book that appeared in 1948, *The Union Challenge and Management Control*, Neil Chamberlain noted labor's increasing interest in the effects of "technological changes," "types of machinery," and "methods of production." He predicted that the "next category of managerial authority in which the unions will seek to deepen and widen their participation will be the category of production" (Chamberlain, 87). Managers resented union interference with their own prerogatives, as they understood them. One executive summed up the issue facing industry in a single question: "Who runs the shop—them or us?" (Noble 30).

As David Noble demonstrates in his recent book, *Forces of Production*, automation commended itself to industrialists, after World War II, precisely because it promised to turn back this threat and to solidify their control over production. The evidence on this point is abundant and unambiguous. According to Earl Troup of General Electric, numerical control of machine production brought "a shift of control to management, [which] is no longer dependent upon the operator." In 1953, a group of engineers at MIT cited among the advantages of numerical control that it eliminated the need for skilled workers. "Little judgment is required and the work is so routine that it is desirable to use a person with little technical skill who will be satisfied with repetitious, entirely prescribed work." A report issued by the Harvard Business School in 1952 pointed out that computerized production meant the "reduction of human attention and skill. . . . Since the

81

control of the machine is automatic, the function of the operator is to load, unload, and start the machine. . . . A skilled machinist is no longer required to operate the machine." *Business Week* reported in 1959 that automated machine tools "run almost untouched by human hands." "The fundamental advantage of numerical control," according to a 1976 editorial in *Iron Age*, is that "it brings production control to the Engineering Department." This consideration led the *American Machinist* to refer to numerical control as not just a "strictly metalworking technique" but a "philosophy of control." A report on automation published by Earl Lundgren in 1969 noted that "a prime interest in each subject company was the transfer of as much planning and control from the shop to the office as possible." Two years later a report issued by the Small Business Administration arrived at the same conclusion. "Much of the skill formerly expected of the machinist operator is now applied by the design engineer, the methods analyst, and the parts programmer." An MIT study made the same finding in 1978. "We believe we see a definite thrust toward deskilling of the N/C machine operators." As a result, "production output, machine downtime, and quality data were more easily obtainable, thus enhancing managerial control" (Noble 232–40).

The drive for managerial control leads to a search for more and more sophisticated machinery and ultimately for the fully automated factory—the managerial paradise held up by *Fortune* magazine in 1946, at the outset of the second Industrial Revolution, as the ideal industrial environment, the "factory of tomorrow" in which human beings have been altogether replaced by machines that "are not subject to any human limitations," "do not mind working around the clock," "never feel hunger or fatigue," "are always satisfied with working conditions," and "never demand higher wages" (Leaver and Brown 165, 204).

Experience with the trouble caused by human beings has occasionally led management in the opposite direction, of course. Instead of attempting to eliminate the "human factor" of production, some companies have delegated managerial tasks to workers in the hope of "motivating" them and of giving them a feeling of participation. Experiments with "job enrichment" and "self-management" have usually been abandoned, however, as soon as managers begin to understand that such programs make many of their own functions obsolete. In Lynn, Massachusetts, a pilot project inaugurated by General Electric in 1968, at the height of the enthusiasm for self-management and for Douglas MacGregor's "Theory Y," was abandoned a few years later, not because the volume or quality of production had suffered but because the union had begun to press for expansion of the program. Like many other managers, those at GE discovered that automation did not free them altogether from dependence on human labor, while its introduction created serious problems of "morale." ("If you treat us like button-pushers," one operator said, "we'll work like button-pushers.") On the other hand, efforts to improve morale by giving workers more responsibility have the unfortunate effect—unfortunate from management's point of view—of whetting the workers' appetite for responsibility and of demonstrating, moreover, that workers are quite capable of exercising it. In the end, the management of GE, according to Noble, sacrificed productivity to power. One executive observed that "productivity may be less real an issue to management than conformity to established work rules" (265–323).

The historical record, then, indicates that industrial technology has grown out of concrete struggles for control over production and takes its existing shape not because this is the shape dictated by ethically neutral considerations of technical efficiency but because it concentrates decision making in a managerial and technical elite. These considerations should make us suspicious of the rosy predictions of a "postindustrial" society in which technological innovation will lead to an abundance of skilled jobs, eliminate disagreeable jobs, and make life easy for everyone. Everything we know about technological "progress" indicates, on the contrary, that it promotes inequality and an unprecedented centralization of political and economic power. Whenever we hear that some new technology is inevitable, we should consult the historical record, which shows that technical innovations usually appeal to industrialists not because they are inevitable or even because they make for greater productive efficiency, but because they consolidate the industrialist's power over the work force. The triumph of industrial technology testifies not to the inexorable march of science but to the defeat of working-class resistance.

It is a muddled, ahistorical view of the industrial revolution that dismisses this resistance as an attempt to "postpone the inevitable," as J. David Bolter writes in his study of the coming "computer age." It is equally muddled to argue that since the "computer age" is upon us, our best hope lies in "reforming the age of computers from within" (Bolter 229). In the past, efforts to reform industrial technology from within, usually led by engineers, served merely to reinforce the lessons already driven home by workers' resistance to the introduction of new technologies: that those technologies serve the interests of capital and that even those who design and manage the machines have little to say about the uses they are put to.

Over and over again, new technologies have reduced even the engineer's work to a routine. What originates as a craft degenerates into a series of automatic operations performed more or less unthinkingly. Computer programming is no exception to this pattern. As Sherry Turkle notes, "Today, programs are written on a kind of assembly line. The professional programmer works as part of a large team and is in touch with only a small part of the problem being worked on" (170). In the early days of the computer, many people hoped that electronic technology could be captured by the counterculture. But things did not turn out that way. Computers encourage centralization and bureaucracy. Instead of humanizing industry, the personal computer came to serve as an escape from industry for hobbyists and even for professional programmers seeking to achieve in the privacy of the home the control they could no longer exercise at work. Turkle reminds us that "people will not change unresponsive government or intellectually deadening work through involvement with their machines at home." But personal computers offer the illusion of control in "one small domain," if not in the larger world of work and politics (175). Sold to the public as a means of access to the new world of post-industrial technology, personal computers in fact provide escape from that world. They satisfy a need for mastery and control denied outlets elsewhere.

There is nothing inherent in computers or in any other type of machinery that leads inevitably to the degradation of work instead of the enhancement of work. Automated machine tools can be used either by craftsmen to perform a variety of tasks or by unskilled operators to perform the same task over and over. In practice, however, managers become uneasy if workers assert too much control over production, and they have accordingly devoted a great deal of energy and imagination to the elimination of the "human factor," as they call it. The ideal machine, in their eyes, is a machine that eliminates the need for human intervention altogether. Insofar as Taylorite principles continue to govern the development of industrial technology, the growing reliance on numerically controlled machines and on computers will simply recapitulate the history of machine tools at an earlier stage in their development. The deskilling of the work force will continue, in other words, checked only by the irreducible dependence on skilled labor that even the most sophisticated machinery has not yet found a way to eliminate.

Sabel and the Thesis of Democratic Technology

But what if Taylorite principles have begun to lose their hold on industry? The hope that the deskilling process can be reversed rests not on the claim that new computer-based technologies will automatically create a new class of skilled workers, as Drucker and other technological optimists assume, but on the demonstration that mass production economies have reached the limit of their capacity to generate sustained growth. Charles F. Sabel, whose qualified optimism about technology distinguishes him both from mindless optimists like Alvin Toffler and John Naisbett and, on the other hand, from critics of technology like David Noble, Harry Braverman, and Lewis Mumford, argues that the "breakup of mass markets" in the seventies and eighties will force American industry to abandon Taylorism and to adopt a more flexible system of production. Taylorism assumes that "large numbers of potential customers have essentially identical and well-defined wants." The specialized markets that are increasingly important today, on the other hand, demand "general-purpose machines and an adaptable work force" (Sabel 201–2). Where critics of technology go wrong, according to Sabel, is in ignoring the decisive influence of markets. It is the market, not some hypothetical struggle for power in the workplace, that determines the uses to which any given technology is put. In a mass market, technology is centralized and hierarchical. When diversified markets prevail, however, technology becomes diversified in its own right: flexible, decentralized, and democratic.

In *The Second Industrial Divide*, Sabel and Michael J. Piore elaborate these arguments in some detail. The saturation of industrial markets, they contend, confronts mature mass-production economies with a choice between two conflicting lines of policy. The first is a "geographic extension of the mass-production system," designed to create new markets for mass-produced commodities and to control the fierce competition that now prevails among the major industrial powers (Piore and Sabel 252). The rise of multinational corporations already foreshadows this policy, but private corporations cannot by themselves bring an integrated global economy into being. Economic integration on this scale, according to Piore and Sabel, would also require a globalization of the Keynesian economic policies that helped to stabilize the national economies of the industrial countries after World War II. An attempt to raise purchasing power would require, among other things, aid to debtor countries and a global war against poverty. Domestically, it would require some form of national economic planning, not to mention an enlargement of the welfare functions of the state. By means of such policies, the United States and other industrial nations might create new markets, in Sabel's words, "without fundamentally shaking management's power to control, invest in, and limit changes in the organization of work" (Sabel 201).

The second line of development points in a different direction, one that is foreshadowed, according to Piore and Sabel, by the revival of small-scale craft production in selected industries in Italy, Germany, and Japan. The second line of development, "flexible specialization," points to a new "technological paradigm" based on short production runs, a highly skilled work force, and a resurgent sense of community. A single example explains why "flexibility depends upon cooperation," according to Piore and Sabel. Under mass production, the training of skilled workers consists of the formal learning provided by the state-supported school system together with a highly specialized training added by the corporation, which enables workers to master particular tasks in a particular firm. Flexible production, however, requires more broadly trained workers, who "can shift rapidly from one job to another." Under flexible production, individual firms have little incentive to invest in training workers who may be hired away from them by competitors; but since all these firms have a collective interest in broadly skilled workers, they find ways to fuse work skills with the "larger life of the community." A system of flexible production thus requires the "regeneration of resources required by the collectivity but not produced by the individual units of which it is composed." These resources include not only the school system but the family, the institutions of local government, and voluntary associations of various kinds. More broadly, they include the tradition of "yeoman democracy" in the United States, as Piore and Sabel call it, as opposed to the market-oriented liberalism that has dominated American politics in the age of mass production. "In market liberalism," they write, "property is to be used to maximal advantage of its possessor; in yeoman democracy, property is to be held in trust for the community. . . . It is this recognition of the indispensability of *community* that makes yeoman democracy . . . the political analogue of the cooperative competition of craft production" (252, 273–75, 299–301, 305–6).

This is an appealing vision of the future, not least because it coincides with a revival of interest, among scholars in a variety of disciplines, in the indigenous tradition of American republicanism on which new forms of economic and political life might draw. The weakness of this analysis lies in its underestimation of the difficulties that lie in the way of communitarian solutions of the crisis of mass production. As one reviewer notes, Piore and Sabel "are strangely silent on how the distribution of economic power dictates the course of technological choice" (Kuttner 31). The present distribution of economic power clearly militates against policies that favor localism, craftsmanship, and the subordination of property rights to local and regional needs. The present distribution of power favors the globalist solution, or even worse, a return to laissez-faire: an attempt to dismantle the regulations imposed on industry in the past, to protect American corporations from foreign competition by means of tariffs and import quotas, and to curtail the power of trade unions. Reaganism, the political face of Taylorism under pressure, amounts to an attempt to counter foreign competition without giving up the technology of mass production.

Technology and the Ideology of Total Control

Piore and Sabel sometimes write as if the demands of flexible, specialized production would in themselves generate a revival of the community life required to sustain it. To show the dependence of flexible production on a reinvigorated civic life is the considerable achievement of their book. But a revival of civic life surely has to be seen as a precondition, not as a consequence of new technologies. The accumulating evidence that mass production economies no longer work very efficiently adds another argument to the already impressive indictment of their dehumanizing effects. The inefficiency of mass production does not in itself guarantee its replacement by a better productive system, however. That depends on a redistribution of power and wealth, on the formation of political movements designed to achieve this, and on a profound change in our values.

Capitalism is more than a system of production for profit, it is also a worldview, an ideology. From the beginning, the technology of mass production was closely bound up with the fantasy that man can free himself from limitations imposed by nature and achieve godlike powers over nature through his own inventions. The dream antedates capitalism, but the productive capacities released by capitalism, together with the scientific revolution, the discovery of the New World, and the Cartesian revolution in philosophy gave it a plausibility, in the modern world, that it had never had before. "In the place of the speculative philosophy taught in the schools we can have a practical philosophy," said Descartes, "by means of which [we can render] ourselves the masters and possessors of nature" (Spragens 56). A Faustian revolt against human limitations, in particular against the human body, held out the hope that intelligence, once it freed itself from the prison of the body—the seat of human limitations—can grasp timeless truths and, through technology, conquer scarcity, sickness, and perhaps some day even death itself. "The day will come," wrote Condorcet in the eighteenth century, "when death will be due only to extraordinary accidents or to the decay of the vital forces, and . . . ultimately the average span between birth and decay will have no assignable value" (Spragens).

Much more than the profit motive, it is this ideology of total control

87

that drives and perpetuates our technology. The determination to eliminate the "human factor of production" is the heart and soul of modern technology, not an incidental by-product of Taylorism that can be discarded now that it has become necessary, we are told, to produce goods for diversified markets. Theoretically, it is possible to design machines that can be used by skilled craftsmen in a variety of ways. Theoretically, it is possible to design a technological system that embodies both a respect for human capacities and an acknowledgment of human limitations. It is theoretically possible to design a technological system, in other words, that satisfies the desire for moral wisdom instead of the desire for domination. But such a system would be completely incompatible with the delusion that underlies the present system, the delusion that we can make ourselves lords of the universe. It would consist of machines designed to enhance and develop human capabilities and to satisfy the instinct of workmanship, as Veblen called it, not to relieve mankind of the need for imagination and ingenuity by assigning these qualities to a small class of technicians. The vision behind our technology assumes that most people find little pleasure in hard work or in strenuous activity of any kind, and it proposes to free them from toil for a life of leisure. It assumes, moreover, that most people are incapable of sustained mental effort in any case, and that even the best minds, indeed, can never altogether free their thoughts from the corrupting influence of emotion and subjective "values."

Here is the most important reason that engineers find it so difficult to entrust their machines to the care of mere human beings. It is not simply the defense of their class privileges that makes them resist demands for worker participation, but the belief that human intervention can only distort and subjectify the beautiful objectivity of the machine. Now that scientists and engineers have devised machines that can allegedly think for themselves, contamination becomes a greater menace than ever before, since the condition of this so-called thinking is precisely that it operate at a level of abstraction where feelings play no part. The promise of the computer age, as it appears to its prophets and propagandists, is the hope that thought can divorce itself from emotion—the most intractable of the human limitations from which technology aspires to deliver us. The utopia of artificial intelligence—the final destination of our civilization, we are told, the earthly paradise that lies beyond even the fully automated factory—rests on the premise that thought can dispense with the thinking self. It can thus overcome the emotional and bodily limitations that have encumbered humanity in the past. Theorists of artificial intelligence celebrate the mind's clarity, as opposed to what one of them, Marvin Minsky, revealingly refers to as the "bloody mess of organic matter" (Turkle 255).

It is often said that modern science implies an assault on human pride, for example by showing that the earth is not the center of the universe, or again by teaching that men are descended from apes. David Bolter restates this cliché when he argues that the computer fosters an awareness of our "temporal limitations" (122). It is more accurate to say that the culture of modern science deplores human limitations but refuses to acknowledge that they are inherent in the human condition, insisting instead that the unaided human intellect can rise above these limitations. The scientific worldview hates the body not merely because it decays but because it is held to be the source of desire. On this view of things, it is because human beings are driven by bodily needs and desires that their understanding is so limited. Only by escaping from these appetites or by overcoming their effects on the consciousness can humans arrive at understanding. The modern faith in disembodied intelligence reaches its climax in the fascination with machines that think with perfect clarity because they have no feelings to get in the way. Here again, what looks at first like a rather disparaging view of humanity barely conceals the grandiose, narcissistic fantasy of annihilating human limitations through the use of machines—the controlling fantasy of modern times, carried here to its logical conclusion. Listen to the feverish speculations of Edward Fredkin of MIT, who once referred to artificial intelligence as the "next step in evolution."

> Basically, the human mind is not most like a god or most like a computer. It's most like the mind of a chimpanzee and most of what's there isn't designed for living in high society [he means advanced industrial society] but for getting along in the jungle or out in the fields. . . . The mere idea that we have to be the best in the universe is kind of far-fetched. . . . The fact is, I think we'll be enormously happier once our niche has limits to it. We won't have to worry about carrying the burden of the universe on our shoulders as we do today. We can enjoy life as human beings without worrying about it. (Turkle, 262–63)

The social vision implied by this kind of thinking is as regressive as the escapist psychology behind it. The psychology is the fantasy of total control, absolute transcendence of the limits imposed on mankind by its lowly origins. As for the social vision, it carries one step further the logic of industrialism, in which the centralization of decision making in an educated elite frees the rest of us from the burden of political participation.

CHRISTOPHER LASCH

Technology is a mirror of society, not a "neutral" force that can "be used for good or evil." It shows us ourselves as we are and as we would like to be; and what it reveals, in the case of the so-called second Industrial Revolution, is an unflattering image of the American at his most incorrigibly escapist, hoping to lose himself—in every sense of the term—in the cool precision of machines that know everything except everything pertaining to that "bloody mess of organic matter."

Works Cited

BOLTER, J. DAVID
1984   *Turing's Man: Western Culture in the Computer Age*. Chapel Hill: University of North Carolina Press.
CHAMBERLAIN, NEIL
1948   *The Union Challenge and Management Control*. New York: Harper.
DRUCKER, PETER
1967   "Technology and Society in the Twentieth Century." In *Technology in Western Civilization*, volume 2. Edited by Melvin Kranzberg and Carroll W. Pursell, Jr. New York: Oxford University Press.
KUTTNER, ROBERT
1985   "The Shape of Things to Come," *The New Republic* 192: 29–32.
LEAVER, E. W., and J. J. BROWN
1946   "Machines without Men," *Fortune* 34: 165 ff.
NOBLE, DAVID F
1984   *Forces of Production: A Social History of Industrial Automation*. New York: Alfred A. Knopf.
PIORE, MICHAEL J., and CHARLES F. SABEL
1984   *The Second Industrial Divide*. New York: Basic Books.
SABEL, CHARLES F.
1982   *Work and Politics: The Division of Labor and Industry*. New York: Cambridge University Press.
SPRAGENS, THOMAS A.
1981   *The Irony of Liberal Reason*. Chicago: University of Chicago Press.
TAYLOR, FREDERICK WINSLOW
1911   *The Principles of Scientific Management*. New York: Harper and Brothers.
TURKLE, SHERRY
1984   *The Second Self: Computers and the Human Spirit*. New York: Simon and Schuster.

6

The Challenge of Economic Transformation
Forerunner of Democracy

BRIAN G. SULLIVAN

All the traditions are worn out, all the creeds abolished; but the new program is not yet ready. . . . Hence what I call the dissolution. This is the cruelest moment in the life of societies. . . .
—Pierre Joseph Proudhon

The signs of economic dissolution abound. Historically high rates of unemployment persist in the midst of a steady economic recovery while factory orders for capital goods continue to recede. At a time when the United States has already displaced two-thirds of the labor force from the production of goods, proposals for improving the American economy embrace a new wave of technology as a means of increasing the rate of industrial productivity.

Since the earliest days of the republic, debate over the economic and social effects of technology has been decided by an overwhelming faith in the promise of technology to benefit all and to minimize side effects of its own making. The latest round of the debate again asks whether the common good, or civic virtue, is compatible with the goal of increased economic efficiency. Yet the central role of technological innovation in mechanizing the workplace has received only scant attention in the economic literature. This should not come as a shock since most economic theorists have built their reputations by explaining how demand, supply and price interact in competitive markets to establish or restore equilibrium.

Technology in Economic Theory

Some economic theorists, both classical and modern, have provided insight into the impact of mechanization on labor. Their insights may help us put the old debate on a new footing. In *An Inquiry into the Nature and Causes of the Wealth of Nations,* Adam Smith indicated a basic problem. He held that efficiency in the generation of wealth is enhanced by the division of labor. But he recognized that specialization involving nothing more than routine tasks diminishes (deskills) the worker by removing all intellectual challenge and responsibility for decision making.

Although David Ricardo, the founder of the classical school of economics, centered his inquiry on the competitive marketplace, he cautioned that increased reliance on mechanization might not be an unqualified blessing. He saw that under certain conditions workers displaced by machines might not be able to secure new employment. It was this spectre of displacement that provoked the Luddites to resist mechanical innovation in 1812. That the Luddite disturbances receded was due in no small part to a demand for skilled workers to operate machines that replaced the demand for unskilled labor.

Karl Marx in *Das Kapital* described not only the economic effects of mechanization but also its adverse affects on workers in mid-nineteenth-century England. Marx contended that the combination of machines, private property, and competition would result in the self-destruction of the capitalist system. In the end, newer and more powerful machines would be introduced, forcing such a large proportion of the labor force out of work that producers would lack consumers to purchase the goods their machines were producing. In the Marxist analysis, new machines are objectionable not because they change society but rather because they leave unchanged the social conflict rooted in an unjust distribution of power between labor and capital. Thus, each new technological change is not value-free because its role in the context of social struggle determines its worth. Each invention counts as a new weapon in the endless struggle between labor and management for social and economic control.

This approach was well documented in Harry Braverman's *Labor and Monopoly Capital.* For Braverman the workplace, under capitalism, is an inescapable battle zone in which workers are required to fight against managers who constantly seek to increase their control over the nature and pace of work and to reduce worker discretion. In a history of the machine-tool industry, David Noble emphasizes that the value of technical innovation depends on how it is applied in the workplace. Because the technology is introduced by management, Noble contends, its goal is to deskill the workers, to leave them powerless while strengthening the hand of management. The traditional focus of study on the marvels of new technology has ignored the cultural transformation that its application requires. Our inquiry, then, concerns applied technology: we must examine not just the machine but how work is organized if we are to assess the social impact of technology.

Fredrick Winslow Taylor's *Principles of Scientific Management* stands as the model of the old industrial order, representing not only the beginning of the managerial era but also signaling the end of the craft era in the United States. Taylor's work amounted to the announced truce, if not the end of hostilities, in an industrial war fought to abolish craft control in American manufacturing enterprises. Initially joined in the Homestead strike of 1892, the struggle pitted the knowledge and skills of craftsmen who controlled the day-to-day manufacturing process against an emerging industrial hierarchy. Battle lines were first drawn over the replacement of skilled craftsmen by foremen and later over the foremen's "autocracy in the shop." This struggle over control, which coincided with the development of large-scale manufacturing enterprises, was the central issue in industrial relations during the era of mobilization between 1890 and 1920.

Taylorism permanently changed the relations between management and labor within the firm. An important condition for the success of Taylorism was the easy availability of replacement labor. Skilled labor was replaced by a horde of cheap, easily trained, and replaceable detail workers. In this system deskilled labor, like the other raw materials, became disposable. When business slowed, labor inventories, too, were inevitably cut.

For the workers, Taylor's cooperative vision simply meant taking orders. Workers' latitude for decision making and control.of their work was drastically reduced. Taylor's system of scientific management installed American commercial Toryism into the core of our industrial organization.

Scientific management made the organization of the large-scale industrial enterprise an extension of the machines at their core. While America's prosperity was, in no small part, a product of the merger between scientific management of the organization and high volume machinery, we must remember Thorstein Veblen's admonition: the way work is organized to suit the requirements of machines determines how men act and dream.

The principles of scientific management were not only adopted by American management but also embraced by labor. American industrial unionism grew up around Taylorism. The emphasis on job definition and control is peculiar to our industrial organization. Labor recognized in Taylor's principles a means for protecting jobs and controlling members. American labor has continued to seek control of the specific details of the labor contract and to protect individual workers on the shop floor through an elaborate system of job definition. Currently, labor's emphasis on job security is a continuation of organized labor's embrace of Taylorism.

Technological Change and America's Regulatory Crisis

We face a fundamental crisis in the United States. The very organization and processes which brought us unprecedented prosperity now threaten our demise. We have begun to recognize our need for new machines and processes but we have failed to renew our organizations in a manner that will provide society with the benefits of the new technology. Our institutional organizations, Tayloristic in nature, cannot maximize the benefits of the new technology either within the organization or throughout society. The principles of scientific management today are as obsolete as Taylor's process for producing high-speed, carbon-tool steel. Both the organization and the machines of production have grown old and both require adaptation.

Robert Boyer and other French theorists describe such a fundamental crisis as a "regulatory" crisis. Regulatory in this sense refers not to the governmental regulations of business but to a fundamental system of institutions that keeps society and the economy in balance (Piore). Thus the fundamental issues that surround the new technologies focus not so much upon the machines themselves as upon the mode of production and distribution within which they are inserted.

The idea that national economies pass through a distinct set of regulatory regimes is fundamental to this theory. Each historical period has its own particular regime with its own set of institutions. These institutions provide stability, coordination, and opportunities for economic growth and development. But with development come fundamental changes in the basic economic structure. These changes eventually render even the successful institutions and regulatory regime obsolete. A new regulatory regime with a new set of institutions is required for the altered economic state.

At the core of America's institutional crisis is an unresolved organizational crisis. The organizational crisis is caused by the breakdown and dissolution of high volume standardized mass production organi-

zations. American economic prosperity depended on large capital in-
tensive industries. These industries combine Taylor's scientific management with large-scale machine production to create long runs of
standardized goods and to achieve efficiencies through economies of
scale. But the institutions that brought prosperity are now obsolete in
the altered economic state.

The central problem for the United States, according to Robert B.
Reich in *The Next American Frontier*, is our inability to move away from
a high-volume mass-production economy toward a more flexible system of production. This flexible system of production is required because of fundamental changes in the world market and in the new
technology. American corporations can no longer compete against the
low labor costs of Third World firms in markets for standardized mass-
produced goods. Reich argues: "The same factors that previously
brought prosperity—the way the nation organizes itself for production—now threaten decline" (119).

Competitive advantage can be regained, in his view, not through
more advanced technology but only through the modification of the
production processes. Advanced industrial economies can meet the
challenge of a globalization of production only in products and processes that demand a skilled work force. These products require custom tailoring, precision manufacturing, or the use of sophisticated
new technologies capable of rapid modification. Such products depend on flexible systems of production.

Reich notes that flexible-production systems depend, first, on the
organization of skilled employees in teams and, second, on the merg-
ing of traditionally separate business functions into a highly integrated system. These systems of production differ fundamentally
from Taylor's scientific management techniques. Flexible-production
systems take advantage of skilled labor in any enterprise where problem solving is more important than routine runs of standard mass-
production products. For modern industrial economies, high-volume
mass-production processes are now obsolete; prosperity can be attained only if innovation replaces standardization. Reich indicates that
in the global marketplace goods are produced wherever they can be
most cheaply produced. Most mass-production industries therefore
locate in Third World countries. Rather than slowing this trend, automation has accelerated it since the new technology reduces the need
for skilled and semiskilled workers in high-volume production.

Despite numerous attempts by American industry, flexible processes of production cannot be grafted upon a business organization
designed for the high-volume mass production of standardized goods.
New managerial techniques prove ineffective because they are grafted

on the old structure of scientific management. It is the underlying structure, not specific techniques, that requires change. The basic principles of old production processes do not apply to flexible systems of production.

In flexible-production systems the process of production is not routine since skill-intensive processes cannot be regimented by a fixed set of rules. The flexible system can adapt quickly to take its advantage in the market only if information is widely shared. According to Reich, "there is no hierarchy to problem solving: solutions may come from anyone, anywhere" (135). While high-volume standardized production organizations are hierarchical in nature, flexible-production systems have few middle line managers and a "flat" structure. Skills of necessity are integrated into a work group. The flexible system of production is organized as a set of stable project teams, with only modest differences in status and income among senior managers and junior employees. But we are not moving in the direction of flexible-production systems fast enough. As Reich points out, "with the close of the management era America has abruptly faced a stark challenge of economic transformation, but the old organization of business, labor and government has resisted change" (119).

Under Taylorism, increased specialization not only forced the development of managerial hierarchies but also pressured American industrial unions to emphasize job control and worker rights on the shop floor. This mass-production model of shop-floor control depends on two key assumptions: a job is a precisely defined series of tasks; and seniority is the criterion for the allocation of jobs. As industrial unionism developed in the United States, each job was sharply defined. A corresponding set of wage and job security provisions was established through the process of collective bargaining. These precise definitions and provisions formed the battle lines in union-management conflicts.

Currently we observe American management concerned with the traditional restraints imposed by the unions to protect workers' jobs. The General Motors–UAW Saturn "experiment" is one manifestation of this concern, but another more common manifestation is managerial avoidance of unionization through the misuse of the National Labor Relations Act. Management argues that these union restraints limit the ability of managers to organize production efficiently, although it should be noted that the desired changes conform to the familiar mass-production model.

American managers feel constrained in their ability to design jobs that utilize the new technologies in a productive way. Labor's limitations seem extreme when compared to the rest of the industrialized world. But unions abroad developed differently. Because national markets were smaller, the standardized mass-production model was not as prevalent. Labor abroad had other protections, including its own political parties, guaranteed lifetime employment and a fuller array of governmental social services.

In these other industrialized nations unions typically negotiate general standards at an industry rather than a corporate level. These standards impose broad limits upon employer practices but leave line managers free to specify the precise details of work. In addition, unions abroad have important policy-making functions at the corporate, industry, and governmental levels which are not yet present in the American system of labor relations. The foreign industrial relations regime substitutes worker participation in policy decision making for shop-floor protectionism. This regime provides workers with protection but it also allows line managers more freedom and flexibility in job assignments.

An alternative to American industrial union shop-floor control exists in the craft model of control. In *The Second Industrial Divide*, Michael J. Piore and Charles F. Sabel suggest two possibilities for the solution of the current international industrial crisis. The first is a "geographic extension of the mass-production system" coupled with a globalization of Keynesian economic policies to help integrate the world economy. The second proposed solution is a departure from mass production toward an industrial economy based upon "flexible specialization." This industrial regime uses short production runs and a highly skilled "craft" work force aided by computer production techniques.

Although current relations and attitudes among business, government, and labor argue for the global solution which leaves unchanged the prevailing processes of mass production and hierarchical management, Piore and Sabel propose an alternative in their analysis of the craft model of shop-floor control. Piore and Sabel examined the craft model of production abroad and the domestic construction trades and the women's-garment industry to sketch an alternative to Tayloristic production regimes.

The craft model provides a different method of protection for the workers. In the craft model wages are not linked to job definition but rather to skill. Job security is not dictated by seniority but rather by limiting access to skills (membership). In construction crafts, only two skill levels are recognized, apprentice and journeyman, and employment security is provided by the hiring-hall system and the careful rationing of skills. Piore and Sabel report that labor and management, under the craft system, are more concerned with substantive resolution than procedural resolution of grievance disputes. In a craft model the work is governed by far fewer procedural rules and substantive outcomes are more important than interpretation of rules and precedents.

Piore and Sabel also examine the American Plan of the 1920s as an alternative to Taylorism. The authors report that the American Plan was a system that attempted to strengthen worker loyalty to industrial firms by offering the worker the opportunity to purchase company stock and participate in various corporate benefit plans. More importantly, the American Plan "provided for a system of dispute resolutions designed to prevent any hardening of the lines between labor and management" (Piore and Sabel 125). Work councils were to resolve disputes: "Their structure and mandate made these councils more concerned with the substantive resolution of production problems than is today's industrial grievance committee" (126). The importance of Piore and Sabel's work is that it gives American labor some alternatives to Taylorism, alternatives which are familiar to American labor.

Now that Taylorism as a regulatory regime and institution has been rendered obsolete, American labor is feeling the imbalance in the underlying industrial structure. American labor has continued, in the face of mounting managerial hostility, to seek control of the specific details on the shop floor through the system of job definition. Tinkering with techniques, management for its part presses for labor concessions without a fundamental change in the regime of relationships.

Worker Participation in a New Industrial Regime

The issue is not economic but political because the changes necessary in the industrial structure are changes in the organization of production not in the underlying economic relationships of supply and demand. If the unions are to modify the shop-floor protections, they must participate in the overall managerial functions. The unions must be allowed to participate in policy decision making.

The times have changed in the forty years since Peter Drucker wrote that "there can be no participation of the worker in the management of the business which in the worker's own interest must be in the hands of trained executives working for the business" (173). The economic future of industrial countries lies in skill-intensive, technically advanced industries which shun hierarchical problem-solving and embrace new products and rapidly modifiable processes to produce them. Without the withering away of the managerial hierarchy, the new industrial regime will be doomed. Reich reminds us that the "radical distinction heretofore drawn between those who plan work

and those who execute it is inappropriate to flexible system production" (135). The new industrial regime cannot succeed without honest
⋯on between labor and capital.

⋯ends on coming to terms with the new technol-
⋯ndustrial organizations to the altered economic
to acknowledge both the threat and the promise
⋯y. Technology's promise is a significant reduction
This promise creates increasing opportunities for
⋯rticipation, education, and artistic creation. The
technology is devastating technological unemploy-
⋯ement leading to economic recession and perhaps to
of capitalism's self-destruction.

⋯nological advance have occurred before. In the 1920s
⋯e through the introduction of electricity and the adop-
⋯rdized mass production. This increased production dra-
⋯ile the total hours of employment were reduced. But the
⋯ge will increase dramatically in the next decade. Comput-
eriza⋯ ⋯astically reduces the need for labor, not only in manufacturing but also in sales and services. In extreme cases, the new technology virtually eliminates the human contribution to production.

In 1983, the Nobel laureate Wassily Leontief published an important
article in *The Population and Development Review*. In it he argued that
while the harnessing of mechanical power in the great Industrial
Revolution eliminated human muscular power as a contribution to
production, none of the "machines could be operated without the
guiding hand, the sharp eyes, and the constant attention of alert
workers, supervised in their turn by skilled foremen" (404). Until recently, capitalism has been creating its own markets, explains Leontief, since its machines have had to rely upon numbers of skilled
workers. But current technological development is altering the machines' need for workers: "one might say that the process by which
progressive introduction of new computerized, automated, and ro-
botized equipment can be expected to reduce the role of labor is similar to the process by which the introduction of tractors and other machinery first reduced and then completely eliminated draft animals in
agriculture" (Leontief 405).

Displacement, Distribution, and Democracy

Leontief indicates that just as tractors replaced farm horses, so too
computerized production is replacing human workers: "the electronic
chip is proving capable of performing more and more complex 'mental' functions that until recently had to be carried out by the human

mind" (405). The danger of the new technology is not only displacement but a shift in the distribution of income and wealth. Middle-
class standards of living decline as the economy shifts away from the
traditional manufacturing base to high-technology industries and the
service sectors. We may create a polarized society divided between a
well-paid elite and a poorly paid mass. The ultimate danger is that
producers may lack consumers to purchase the goods their machines
will produce.

To face rather than ignore both the promise and the threat of the
new technologies, business, government, and labor must develop a
new regulatory regime of production and distribution. This new regime requires an increase in participation not only in the process of
production, but also in the dynamics of distribution. The danger here
is that production work is being increasingly automated just at the
time when the agents of distribution, the public sector and the trade
union movement, are in serious decline.

Piecemeal reforms will continue to prove inadequate. Leontief argues that the loss of purchasing power from the decline in wages
brought on by automation could doom the economy to rising productivity and rising destitution. He proposes a shorter work week to ensure a fair distribution of jobs and a greater reliance on government
transfer payments to distribute income and maintain consumer purchasing power. A democratized production and distribution system is
essential if American society is to benefit from the introduction of the
new technological system.

In this atmosphere of dissolution, a renewed trade union movement can contribute greatly to the stability and coordination of the
new regulatory regime. The new model of American industrial relations must provide for broader job classifications, new employment
security provisions to include codetermination, income and employment guarantees, training programs and a redefinition of the union's
role in protecting worker interests.

In *What Do Unions Do?* Richard Freeman and James Medoff claim
that "in the economic sphere, unions reduce wages inequality, increase industrial democracy, and often raise productivity, while in the
political sphere, unions are an important voice for some of our society's weakest and most vulnerable groups, as well as for their own
members" (5). Freeman and Medoff recognize that unions have two
faces: "a *monopoly* face associated with their monopolistic power; and
a *collective voice/institutional response* face associated with their representation of organized workers within enterprises" (6).

Traditional union action will gain little if organized labor continues
to pursue objectives associated with its monopoly face because a mo-

nopolist loses power when an effective substitute for a product he sells comes onto the market. Modern trade unionism and the new industrial regime will benefit only from labor strengthening its collective voice. By providing communication for workers as a group with management in large enterprises, trade union's exercise their collective voice. Freeman and Medoff point out that collective rather than individual bargaining is required for two reasons. First, many aspects in an industrial setting involve the common good of workers, and second, the union's collective voice protects workers from managerial retaliation. The possibilities for significant change begin when workers' participation in decision making is required by the new production system. Then, as Daniel Zwerdling observed, "even the most superficial labor-management projects contain within them the seeds of more important changes as they awaken workers to a process of thinking, discussing and challenging" (19).

The case for workers' participation in policy making within corporations has been strengthened by Robert A. Dahl's *Preface to Economic Democracy*. Dahl contends that major inequalities in American wealth and income are caused primarily by "a highly concentrated ownership of property and very large payments to top corporate executives" (139). He argues further that a system of self-governing enterprises would enable Americans to enjoy a greater measure of distributive justice than is currently possible under our present system of corporate capitalism.

Corporate America, Dahl notes, has adopted the view that "property is the superior, self-government—the subordinate right" (163). This development has given legality and legitimacy to a "system of government Americans view as intolerable in governing the state, but has come to be accepted as desirable in governing economic enterprises" (162). Authoritarian leadership, typical of large American corporations, is ill-suited to innovation and growth since it cuts leaders off from their own work force. Conversely, small firms, because they are self-governing, could confront the challenge of innovation successfully.

In his *Principles of Political Economy*, John Stuart Mill delivered a sobering commentary on the hope for a more humane, democratic work place. After remarking that Americans were a race in which the "life of the whole of one sex is devoted to dollar-hunting, and the other to breeding dollar-hunters," Mill observed that the "best state for human nature is that in which, while no one is poor, no one desires to be richer, nor has any reason to fear being thrust back, by the efforts of others to push themselves forward" (337). Mill pointed out that an increase in production is only important in backward countries; the more advanced countries require better distribution. He concludes: "It is questionable if all the mechanical inventions yet made have lightened day's toil of any human being. They have enabled a greater population to live the same life of drudgery and imprisonment, and an increased number of manufacturers and others to make fortunes" (340).

What is necessary for the transformation of regulatory regimes? Fundamentally, a recognition of both the problems and the promises of the new technology and a commitment to societal transformation, including reappraisal of the institutions underlying corporate capitalism. A reappraisal of such proportions was avoided even in the depths of the Great Depression but without it both our society and our economic organizations will stagnate.

Works Cited

Braverman, Harry
1975   *Labor and Monopoly Capital: The Degradation of Work in the Twentieth Century.* New York: Monthly Review Press.
Dahl, Robert A.
1985   *A Preface to Economic Democracy.* Berkeley: University of California Press.
Drucker, Peter
1946   *The Concept of the Corporation.* New York: Mentor.
Freeman, Richard B., and James L. Medoff
1985   *What Do Unions Do?* New York: Basic Books.
Leontief, Wassily
1983   "Technological Advance, Economic Growth and the Distribution of Income," *Population and Development Review* 9: 403–10.
Marx, Karl
1984   *Capital.* Edited by Friedrich Engels. Chicago: Encyclopedia Brittanica.
Mill, John Stuart
1923   *Principles of Political Economy.* New York: D. Appleton and Co.
Noble, David F.
1984   *Forces of Production: A Social History of Machine Tool Automation.* New York: Alfred A. Knopf.
Piore, Michael J.
1982   "American Labor and the Industrial Crisis," *Challenge* 25: 5–11.
Piore, Michael J., and Charles F. Sabel
1985   *The Second Industrial Divide.* New York: Basic Books.
Reich, Robert B.
1983   *The Next American Frontier.* New York: Times Books.

RICARDO, DAVID
1974   *Principles of Political Economy and Taxation*. New York: Dutton.
SMITH, ADAM
1984   *An Inquiry into the Nature and Causes of the Wealth of Nations*.
               Chicago: Encyclopedia Britannica.
TAYLOR, FREDERICK WINSLOW
1967   *Principles of Scientific Management*. New York: Norton.
VEBLEN, THORSTEIN
1973   *The Theory of the Leisure Class*. Boston: Houghton Mifflin.
ZWERDLING, DANIEL
1978   "Workplace Democracy: A Strategy for Survival," *The Pro-
               gressive* 42: 16–24.

7

Industrial Technology and Collective Bargaining

IRVING BLUESTONE

In September 1984 the Eurojobs Conference, held in Paris, emphasized two basic themes:

1. Compelling global competition, accent on quality improvement, a better educated work force, external environmental pressures, and the rapid advent of new technologies all require management to embrace affirmative and even aggressive change, in corporate culture, industrial relations, and the traditional view of the role of business in society.

2. In employer-employee relations, the past is not a guide to the future; the question is not how to fit the workers to the technology, but rather how to fit the technology to the workers.

The conference called attention to the speed of change and the urgent need to recognize that human considerations should take precedence when introducing advanced technology. Such a conclusion has its roots in moral and ethical values. In equal measure, however, it derives from a clear understanding of hard-headed, economic benefit.

The Problem of Structural Unemployment

Recent national, social, and economic crises also call for scrutiny of the "human face of technology." The upward drift of unemployment rates in the years following World War II serves as a strong signal that structural, aside from cyclical, unemployment is a serious problem perhaps too long ignored. Consider for a moment these findings in a study undertaken at the University of California (Brown 177). The average rate of unemployment for each national administration in the last forty years indicates:

Truman	4.2%
Eisenhower	4.9%
Kennedy/Johnson	4.9%
Nixon/Ford	5.8%
Carter	6.5%
Reagan (1981–83)	8.9%

The average unemployment rate for the the 1981–85 period of the Reagan administration was 8.3 percent, a sharper increase than any of the noted preceding administrations.

Deindustrialization has played a major role in this turn of events. As technology advances into the twenty-first century, this alarming trend will continue unabated unless society, business, labor, and government are alert to the dangers which structural unemployment promotes and adopt policies and programs to bend the trendline downward.

Who Is Affected?

In late 1982, the president of Carnegie-Mellon University, Richard Cyert, in a report on the future of technological advance, concluded that robots could replace one million workers by 1990 in the automotive, electrical equipment, machinery, and fabricated metal industries alone and three million by the year 2000. And "sometime after 1990, robot capabilities will be such as to make all (7.9 million) manufacturing operatives replaceable" (*BNA Daily Report* A.3). Other reports on the subject have arrived at similar conclusions, although prognostications concerning the depth of decline in the manufacturing work force vary.

The April 25, 1983, issue of *Business Week* examined at length the influence of new technology on the declining number of middle managers and the growing threat to both their usefulness and job security. The article's captions alone effectively describe the grave uncertainties surrounding middle management, directly attributable to "pains and strains" of the revolution in electronics: "The shrinking of middle management"; "Who will retrain the obsolete managers?" "The disenchantment of the middle class"; "Companies will be leaner, more fluid, with fewer levels of management . . ."; "The old foreman is on the way out . . ."; and the painful statement by a redundant manager, "all of a sudden you don't have to be anywhere" ("Special Report" 50–86).

At all levels, from the traditional typist to the engineer, from the machine operator to the skilled maintenance crew—in factory, office, and store—technological breakthroughs will erode job opportunity. It is assumed that nonmanufacturing employment will continue to rise, and that shifts within the work force therefore will limit overall employment disruptions. However, the nonmanufacturing sector will experience equal or even more rapid changes than those experienced by blue-collar workers. The problems to be met, therefore, are similar for both the manufacturing and nonmanufacturing sectors. The spotlight is not just on the blue-collar worker.

There is, of course, an ongoing dispute whether the work force made redundant by the introduction of electronically controlled equipment and microprocessors will be absorbed as new industries arise. Whether or not this forecast proves true, untold numbers—ranging from nonskilled blue-collar workers to the upper reaches of management—will unquestionably suffer.

The purpose of this chapter is to highlight those issues which fall within the collective bargaining arena. Anticipating fast-moving change and planning for solutions to the myriad problems which will arise as a result of technological progress are essential rational requirements of a compassionate society. In the arena of labor-management relations, union leaders and managers must undertake joint action to find ways in which they can better manage accelerating initiatives in technology.

It is customary for management and the media to paint a bleak picture of union resistance to technological change. The popular perception, in large measure stimulated by the media, is that unions insist on preserving the technological status quo, displaying an insensitivity to the continuing need for innovation "to keep up with competition." I might add, on the flipside of the coin, that business is commonly perceived as panting to bring new technology aboard, when, in fact, economic history is replete with examples of creativity being suppressed by business interests during periods of technological advance. Two examples should suffice. After the oxygen process for making steel was developed in Austria, U.S. firms were very slow to adopt the process. In fact, foreign competitors still have an overall technological advantage in this area. Steel-belted radial tires are another example of delayed adoption of a technological breakthrough by American corporations.

While labor resistance to change has occurred from time to time (the struggle to maintain firemen in diesel locomotives was the subject of widespread media coverage), the fact is that new technology generally has not only been accepted by labor but has been embraced as one of several means to create a larger economic pie, to remain competitive, and to provide the opportunity for improving the standard of living of the nation's families. Thus almost thirty years ago, at a time when the term *automation* was on everyone's lips, Walter Reuther, then president of the International Union, United Auto Workers, declared:

> Labor . . . is not opposed to automation. We fully recognize the desirability, as well as the inevitability, of technological progress. We welcome the potential benefits which automation could and should bring. But we oppose those who would introduce automation blindly and irresponsibly, with no concern for any result except the achievement of the largest quick profit for themselves. (Reuther 75)

As early as 1948, the labor contract between the UAW and General Motors Corporation contained a provision which gave clear recognition to the fact that "a continuing improvement in the standard of living of employees depends upon technological progress, better tools, methods, processes and equipment, and a cooperative attitude on the part of all parties in such progress" (Par. 101b).

In more recent years, the negotiating parties began to grapple with the problem of technology and job erosion by offering practical approaches to possible solutions. The answers may be found in management's acceptance of the concept that the benefits associated with the introduction of new technology are shared equitably with the work force and that the employee is viewed not as a "hired hand" to be discarded—the customary practice—but rather as a resource to be retained and developed. Such an attitude on the part of management and labor presents the opportunity for meaningful negotiation over issues related to the human consequences of new technology.

A prerequisite to success requires a departure from usual collective bargaining practices in which management decides on a course of action and the union responds by challenging the action taken. This practice compels "after the fact" controversy rather than "before the fact" planning, designing, and reaching agreement. Let us examine some of the problems which the parties should address and the possible negotiated understandings which might satisfy the human concerns of the work force and the reasonable requirements of the business.

IRVING BLUESTONE

Layoff Avoidance

Beyond question, the most provocative issue confronting the parties in solving the problems attending technology advance is employment security. It was the fear of permanent job loss that provoked the Luddite movement. The same concern permeates the current scene. Labor contracts contain intricate, comprehensive seniority rules which determine who will keep or lose their jobs, but they do not usually provide for layoff avoidance. The rule has been for the union to monitor violations of seniority rights and respond with grievances if an employee is improperly laid off. In recent years, however, the concept of layoff avoidance, the retention of employees rather that their release, has come front and center at the bargaining table.

To clarify the issues and the process of collective bargaining approach, it might be helpful to review the historical development of a specific situation. The statement of principle, noted earlier, that was incorporated for the first time in the 1948 UAW-GM National Agreement recognized the desirability of sharing the economic benefits of technological progress by improving the employees' standard of living. It was not until later years, however, that serious attention was directed toward the issue of protection against job loss. In 1967 the UAW and GM agreed to establish a joint National Committee on Technological Progress as an initial response to the introduction of computer-controlled equipment and the annoying management practice of placing new technology functions outside of the bargaining unit, thereby reducing the number of jobs represented by the union. Emphasis was squarely placed on the joint committee's involvement in helping to resolve disputes "occasioned by advancing technology, causing a shift of work from represented to non-represented employees" (*UAW/GM National Agreement* 1967, 12). By reason of the work of the committee, strengthened by precedent-setting arbitration rulings, work was preserved in the bargaining unit and jobs were thereby saved for affected employees.

In the 1979 negotiations the parties wrote new language which required that local management give advance notice to the local bargaining committee whenever it wished to introduce new technology. This provision permitted discussion and agreement at the local level concerning the impact of the technology on the work force; it also provided for referral to the national parties in case of local failure to solve the problems ("Statement on Technological Progress" 375–78). Almost three years later, provisions were negotiated in the historic 1982 agreement dealing further with the employment security issue. Several unique notions were adopted, including the establishment of a corporation financed fund to train or retrain employees to save them from being permanently released ("Joint Skill Development and Training Program" 277).

The 1984 national contract once again placed special emphasis on the concern to avoid layoffs. Indeed it was the highest priority issue in the negotiations, announced by the declaration unanimously approved by the delegates to the UAW's Special Collective Bargaining Convention held in March 1984. An excerpt from the comprehensive statement entitled "Employment Security" serves to clarify the concept adopted by the union:

> An essential part of developing . . . employment guarantees . . . will have to involve the employer's commitment to actively seek ways to utilize the skills and talents of bargaining unit members. Instead of viewing them as dispensable parts of the business, and—as often seems to be the case—looking for ways to get rid of workers, management must work with us to assure that bargaining unit members can do whatever work needs to be done and that they have a fair opportunity to fill available jobs. (23)

The 1984 agreement introduces imaginative, innovative ideas which serve as a sound foundation for the full development of layoff avoidance programs. One can assume with reasonable confidence that future negotiated provisions will build on this foundation. The centerpiece of the program is the Job Opportunity Bank-Security (JOB Security) Program. It provides that no employee with one or more years of seniority will be laid off as a result of the introduction of new technology, outsourcing, negotiated productivity improvements, or due to job loss attributable to the movement of work to another GM plant, or the consolidation of component production. The program is underwritten by a financial commitment of more than $1 billion until the end of the succeeding national contract. Most noteworthy, the entire program will be under the joint control of UAW-GM JOB Security Committees at the local, area and national levels. Emphasis is on the equality of the parties at all levels in implementing the program ("Job Opportunity Bank-Security" 23).

Another innovative feature of the agreement provides for the creation of a joint Growth and Opportunity Committee whose purpose is to develop and launch new business ventures aimed at providing job opportunities for union represented employees. The program is supported by a fund of up to $100 million committed by GM. At the local level joint committees are charged with developing ideas for business ventures which will be reviewed for feasibility and implementation by the national committee ("Job Opportunity Bank-Security" 299).

Among several additional provisions aimed at creating enhanced security are improved training and self-development programs and commitments by the corporation to proceed with new small-car production. The much-publicized agreement concerning the small car called Saturn was reached between the parties in July 1985 in fulfillment of that commitment. Sharply increased early retirement pension benefits are intended to spur job openings as well as afford more leisure time over the working life of employees.

Once the negotiating parties are committed to finding their way together toward enhanced employment security for the employees, a broad array of mutually agreeable responses becomes possible. Citing some of them should provide an understanding of the opportunities available for sensible, rational solutions.

Prenotification

The introduction of new equipment should be preceded by sufficient advance notification to the union so that negotiating parties have an opportunity for constructive discussion. Many labor contracts already contain prenotification provisions as an essential element in paving the way for sensible management of obligations to the affected work force.

Formation of a Joint Committee on Technological Progress

As a measure of management's commitment to sharing with the union its contemplated technological changes, it is advisable to establish a joint committee on technological progress. Such a committee would meet regularly to discuss the development of new technology and its impact upon various aspects of the collective bargaining relationship. The committee would receive advance notification as noted above. Armed with the necessary information, it could conduct more targeted discussions concerning any resulting problems.

What, then, are some of the problems which naturally arise as new technological equipment is introduced? Obviously, layoff avoidance lies uppermost in mind. Others, most of which are related to employment security, fall into several categories.

Maintaining the Integrity of the Bargaining Unit

It is not unusual for management to place technologically advanced equipment outside of the bargaining unit, even though it is merely a substitute means of performing the same functions previously assigned to bargaining unit employees. It is self-evident that the union will object strenuously to such a decision and insist that the new equipment remain within the bargaining unit. Early discussion, complemented by experience in training bargaining unit employees to handle such new operations, should avoid irreconcilable confrontation on the issue. Nevertheless, it is an area for controversy, especially if management is deliberately attempting to withdraw jobs covered by the bargaining unit, providing jobs for nonunit employees at the expense of represented employees.

Training and Retraining

Essential to a rational introduction of new technology is the obligation to provide adequate training to those affected so that they are capable of efficient and quality performance. One of the shortcomings of U.S. management has been the failure or refusal to undertake the costs involved in providing extensive training. Training should be comprehensive and varied. It also should excite interest in the product or service by showing how respective functions fit into the larger production picture and how the product or service plays a meaningful role in society.

Economic Issues Related to New Technology

Management often attempts to alter classification titles and accompanying wage rates to accommodate the new technology. This, of course, is a matter for collective bargaining and can become a sticky issue. Often enough, management argues that the technology has eliminated skills and provides more comfortable conditions of work— the basis, they claim, for reducing wage rates. The union will resist such efforts, arguing that the purpose of the "new" operation is no different from the previous production method. On the other hand, the new technology may require an upgrading of skills and knowledge for which additional training is needed. The argument then becomes reversed. It depends largely on whose ox is being gored. Nevertheless, by and large, these seemingly highly controversial issues tend to work themselves out. They are, however, tinderboxes which, in major shifts of the production process, can inflame the relations between the parties.

Other issues of an economic nature stem from decisions to reduce manpower as a direct result of technological progress. Various devices are utilized to cushion the blow or avert it altogether. Supplementary unemployment benefits, designed to provide economic relief from a layoff, are also applied to layoffs resulting from the introduction of advanced technology. Similarly, severance pay programs come into play, particularly when the layoff is permanent. Efforts are made to invoke special early retirement pension benefits, keyed to the unique circumstance of technologically induced redundancy. In some instances, provision is made to transfer such laid-off employees to other facilities of the corporation, using geographic area hiring preference procedures or even nationwide hiring rights in multiplant operations. And, of course, the drive to reduce work time to create more job opportunities is high on the list.

Health and Safety Issues

The introduction of new and different types of advanced technology requires a careful monitoring and analysis of resulting health and safety hazards. This applies both to blue collar and white collar functions. The negotiating parties would do well to establish a joint committee—aside from the work level health and safety representatives—that would have continuous access to comprehensive information concerning the nature of the new technology (including the use of chemical substances), would undertake special research into possible hazards (either in-house if appropriate capability is available or through expert consultants), and would make available the findings whether benign or malevolent. Financing for such studies could be managed by creating a fund, either lump sum or on the basis of "cents per hour" supplied by the company.

Technology and Employee Surveillance

Technological developments are sometimes viewed by management as an additional tool to monitor the behavior of employees, to keep track of production, quality, down time, relief time, etc. Various devices are used as "Big Brother" surveillance monitors to provide evidence for purposes of disciplinary action. Such uses result in low employee morale, resentment, resistance and strategies to "beat the system." Such devices become counterproductive in the same way that autocratic behavior by management in the treatment of employees eventually backfires. There are certain legitimate uses of electronic technology to maintain production and good quality. There should be clear understanding and agreement, however, that these devices are not to be utilized for purposes of surveillance, disciplinary action, invasion of privacy or control over individual performance.

The Quality of Work Life Improvement Approach

Experience indicates that some of the problems attending technology advance are either blunted or altogether resolved where a successful Quality of Work Life (QWL) process is in effect. For instance, in anticipation of the installation of new equipment, the employees who will be assigned to it may visit the worksite where the equipment is being produced, not only in order to become acquainted with it but also to offer suggestions and advice directed toward comfort, design or improved operational handling of the equipment. Employees who have the opportunity to exercise meaningful control over the methods, means and processes of manufacture or service are more prone to adopt new equipment as their own and take greater pride in its purposes and functions—provided it does not represent a direct threat to their own employment security. The concept of QWL is designed to bring to the workplace a new system where managerial attitudes toward the employees, the adult human resources, evolve from the orthodox rigidity of "scientific management" to a clear recognition of the dignity of adult human beings and their innate capacity for creativity and self-fulfillment.

Experience with the QWL process over the past fifteen years of implementation has proven its value far beyond the arena of technology introduction. However, its success in improving life at work for employees, which is its immediate goal, depends greatly on the genuine commitment by union and management alike to its day-to-day application. In fact, where the negotiating parties clearly understand and appreciate the need for change in work organization and engage in mutually desirable action programs aimed at democratizing the workplace, they strengthen the collective bargaining process. The essence of collective bargaining as social policy is to resolve employer-employee disagreements. The QWL process is not designed to eliminate controversy. Tough bargaining continues over controversial issues. QWL, however, is an appropriate vehicle for the resolution of an array of issues which the parties recognize as mutual concerns.

Collective bargaining serves as an effective democratic method of solving workplace problems. The expansion of its scope and breadth into a relationship in which labor and management jointly seek solution to mutually recognized problems paves the way for ready acceptance by employees of the need for technological innovation while alerting management to the need for humane solutions to problems created by the new technological society. In the final analysis, society must reach the conclusion that the fate of a nation lies not just in the technology of its machines but in the vitality of its citizens.

Works Cited

1982	*BNA Daily Report* (Nov. 1): A.3.

BROWN, CLAIRE

1983	"Unemployment Theory and Policy: 1946–1980," *Industrial Relations* 22: 164–85.
1984	"Employment Security—Resolution," UAW 1984 Special Convention—Collective Bargaining Program.
1984	"Job Opportunity Bank—Security," *UAW-GM National Agreement.*
1982	"Joint Skill Development and Training Program," *UAW-GM National Agreement.*

REUTHER, WALTER

1961	*Selected Papers.* Edited by Henry M. Christman. New York: Macmillan.
1983	"Special Report: A New Era for Management," *Business Week* no. 2787: 50–86.
1979	"Statement on Technological Progress," *UAW-GM National Agreement.*
1948	*UAW-GM National Agreement.*
1967	*UAW-GM National Agreement.*

PART THREE

Community

8

Medical Technologies and Communities of Moral Value

PAUL F. CAMENISCH

Technology not only reshapes the impersonal world of things, it inevitably affects persons and communities and alters our ways of perceiving and ordering the human world. Thus people have often acted together to resist or to redirect technologies which they perceived as potentially harmful. But a purely negative stance aimed only at avoiding certain palpable and almost universally feared ills—injury, death, intolerable costs, etc.—does not make such groups communities. Community, as used here, refers to persons united by certain shared values or beliefs which significantly shape the relations among them in enduring and positive ways (Taylor 26). Such guiding values or beliefs go beyond the harms to be avoided and can enable the community to set limits to technology and even to redirect it in creative ways.

The last fifty years of medicine provide us with three conspicuous examples of this sort of community-technology interaction, involving three kinds of communities: a religious one, Jehovah's Witnesses; a professional one, physicians; and a less traditional one, a sharply focused voluntary association, the hospice movement. Both the differences and the commonalities among these three quite different examples will help elucidate this potentially positive form of response to modern technology.

117

PAUL F. CAMENISCH

Jehovah's Witnesses and Blood Transfusion

Founded in 1872 under Charles Taze Russell, renamed Jehovah's Witnesses in 1931, and currently claiming over two million members worldwide (Clarke 497), Jehovah's Witnesses are of interest here because of their position that blood transfusions, including administration of packed red blood cells, plasma, white blood cells, and platelets, are equivalent to the biblically proscribed eating of blood (Genesis 9:3–4; Lev. 17:11, 14; Acts 28:29; Dixon and Smalley 2471). While they had long put the highest value on adherence to scripture's teachings, and had exalted fidelity to their religious beliefs over all other considerations, the specific teaching concerning the medical use of blood emerged only in 1945 (Bonakdar et al. 587).

That belief set the stage for a lonely battle against a promising medical technology, a battle made more lonely by the other ways—religious, moral and political—in which the Witnesses were marginalized or alienated from the society.

Abstaining from blood is related to the Witnesses' belief that the blood contains the very life of a being. Violation of this prohibition endangers the believer's chance of resurrection and of eternal salvation. However, it is incorrect to assert that their "objection is purely doctrinal," for they also cite the potential physical hazards of blood transfusions such as the transmission of disease and various complications arising from the complex matter of matching blood types (Clarke 498–99; Ott and Cooley 1258). In fact, 40 percent of their major booklet on blood, *Blood, Medicine, and the Law of God,* is devoted to possible medical objections to transfusions (Farr 41). While not mandatory for membership, this position is widely accepted among Jehovah's Witnesses. One survey of a congregation has revealed unanimous opposition to blood transfusions among adult members, and high percentages as having refused blood transfusions (29 percent), or having had "bloodless surgery" (46 percent) (Findley and Redstone).

The impact of this position on the health care community has been complex, involving the attitudinal response of health-care deliverers, legal liability, institutional and personal treatment policy, more careful reassessment of the technology involved, and development of alternative technologies. The attitudinal responses of health care deliverers have ranged from annoyance at such "interference" in medical judgment and practice (Dixon and Smalley 2472), through a conflict between the values of providing the best medical care possible and the need to respect the wishes of the patient (Dornette 274), to an insistent drive to develop techniques which will serve both those values and so enable the health-care deliverer to treat the "whole person" of the patient (Dixon and Smalley 2472). The Witnesses' position has prompted some institutions and individual practitioners to refuse to treat them because of the greater risks involved (Dixon and Smalley; Gonzalez 720).

The Jehovah's Witnesses' stand has generated pressure on the medical community for the continuing assessment and pursuit of alternatives to blood transfusions. While many physicians predicted frequent dire consequences from refusal of blood transfusions, a more careful examination of the professional literature confirms that there were also possible dire consequences to accepting blood transfusions ("Foreign Letters"; Brewster et al.; Gonzalez); that the consequences of refusing were not frequently as serious as had been anticipated; that the use of transfusions had in many settings become routine, even casual; and that preferable alternative therapies were available for many problems.

Major gynecologic, obstetric, and cardiovascular surgery is now done virtually routinely on persons refusing transfusions (Bonakdar et al.; Ott and Cooley). Dr. Ron Lapic, frequently relying on electrocautery, reports only one death related to transfusion refusal among his approximately 2,500 Jehovah's Witnesses' patients (Gonzalez 720). By 1981 Cooley, on the basis of 1,026 operations, stated that "the risk of surgery in patients of the Jehovah's Witnesses' group has not been substantially higher than for others" (Dixon and Smalley 2471). However, deaths related to transfusion refusal are still reported (Harris et al.), and certain complications of pregnancy in such situations remain a dreaded problem for practitioners (Bonakdar et al. 590).

The evidence concerning transfusion's persistent risks and the success of surgery without it indicates that the Witnesses' stand is neither as irrational nor as life-threatening as many thought it would be. However, others have also benefited from their courageous stand. The pressure they exerted on conscientious health-care deliverers to provide quality care while respecting the whole person of the patient played a significant role in the development of alternative techniques and technologies for dealing with blood loss, developments from which the general population stands to benefit.

The major categories of alternative techniques are "blood extenders," autotransfusion, and "bloodless" surgery. Reasonably successful "blood extenders," which are actually only plasma volume expanders, include Dextran, a "preparation of depolymerised polysaccharide" (Farr 35–36; cf. Ott and Cooley 1256), and Fluosol-DA, the first three American recipients of which were Jehovah's Witnesses (Gonzalez 720). Greater use of autotransfusion also circumvents the objections to and most of the risks of transfusions. However, Jehovah's Witnesses accept only blood which can be suctioned, filtered, and returned immediately to the cardiovascular system during surgery, since blood which has lost contact with the living body is unacceptable to them. Reports indicate that this procedure need not lead to problems such as contamination or coagulation abnormalities (Brewster et al.). "Bloodless surgery" is of course a relative term referring to surgery undertaken with an eye to minimizing blood loss. The use of electrocautery not only circumvents objections to and risks of blood transfusions, but generates other benefits resulting from the increased ease and speed of surgery made possible by a "nearly bloodless operative field" (Schaller et al.). Other relevant techniques include the administering of hematinics to elevate the hemoglobin level to compensate for blood loss (Smith 659) and the use of hypotensive anesthesia to decrease blood pressure and thus to slow blood loss (Gonzalez 720).

The Jehovah's Witnesses' position also drove physicians to consider more carefully the larger moral dimensions of what they were doing, dimensions which emerged when practitioners had to acknowledge the fact that they were here dealing with a socially powerless group (Gonzalez 724) which it would be all too easy to abandon or to try to intimidate. Out of that acknowledgment came a renewed commitment on the part of many practitioners to treating the whole person, to ministering to the physical needs in a way that did not conflict with the patient's spiritual and moral beliefs (Bonakdar et al. 589; Clarke 498; Dixon and Smalley 472). Thus a community taking a deviant position on an accepted medical technology, a position initially frustrating to the moral-professional community of medicine, in the end challenged and enabled the latter to be true to its own moral commitments of caring for the whole patient. Ensuing technological improvements enabled both communities involved to remain true to their values and commitments: the Jehovah's Witnesses to their religiously based rejection of blood transfusions and their valuing of life; and the medical community to its dual commitments to preserve life and health, and to serve the whole patient.

Hospice and the Technology of Dying

Not all will readily agree that hospice and its relation to the medical technology of dying—those numerous life-prolonging means which can imperceptibly become death-prolonging ones, those "curative" measures frequently continued after all hope of cure is gone—exhibit the creative interaction between a community and a technology we have seen above. Some will see hospice as simply another health-care institution, a different delivery mode for professional services, and not as a community. And others, following a persistent caricature of hospice, will see its response to medical technology not as a creative limiting and redirecting, but as a disdainful dismissal of a well-intentioned, but misguided project. The task of the following section is to show that these responses are incorrect, that indeed the hospice-technology encounter does conform to the pattern found in the case of Jehovah's Witnesses and blood transfusions.

Even though these two examples reflect the same general pattern, the encounter between hospice and the medical technology of dying involves some dynamics quite different from those just treated. Here the offending technology itself generates the resisting community, a community which is not an explicitly religious community, but one whose most fundamental motivating beliefs, unlike those of the Jehovah's Witnesses, could be made comprehensible, perhaps even persuasive to the larger society and its medical establishment. The very *raison d'être* of hospice is its discomfort with prevalent medical technology and its desire for a more humane alternative. Therefore, the hospice movement's approach to the technology was much more aggressive and creative than was that of the Witnesses, who, except for occasional suggestions of alternative therapies (Farr 35–36), were content to reject the technology and quietly to pay the price of that rejection.

The hospice phenomenon, springing in its current form from the work of Cicely Saunders of St. Christopher's Hospice in London (Munley 28ff.), has been a major element in the contemporary response to death. A 1980 survey identified over 800 hospice programs in operation or in some stage of development in the United States. In 1974 there had been none (Smith and Granbois 8).

The major stimulus for this phenomenal growth appears to have been a reaction to the complex medical technology that had come to dominate the experience of death and dying which was increasingly occurring in the hospitals. Madalon Amenta, a nurse and hospice activist, reflects this dimension of hospice when she describes "the psychosocial conditions produced by the dominance of cure-oriented medical technology in the medical care system and the reaction to them that hospice programs have come to express" (1985, 279). The urgency of this stimulus is reflected in the impassioned cry of one who watched his wife's prolonged dying: "It was a fetish, nothing less, for society to worship the flesh while it destroyed the spirit . . . let us rise, all of us, to defend the defenseless body against the machine" (Halper 18).

The technological extension of dying was seen not only as unproductively adding to patients' and families' sufferings, but as leading to "a premature psychological death" because of the patient's exhausted and drugged confusion often resulting from such prolonged interventions. According to many hospice supporters, the social and psychological abandonment naturally feared by the dying, is simply aggravated in the hospital setting by the greater difficulties of access by family and friends (Corbett and Hai 39ff.), and by the distance often created between patient and medical professional by the intervening technologies (Halper 17; Reiser).

Hospice offered a "radically different milieu for dying" (Corbett and Hai 40–41; cf. Munley 27; Krant 1062). This attempt "to reclaim dying from the technological, nonhuman order and restore it to the moral, human order where it belongs" (Munley 5) involved both general commitments and specific practices. Among the former were realistic recognition of the inevitability of death and of medicine's and medical technology's limited ability to forestall it; a recognition of the crucial importance of responding to the whole person of the dying one, including that person's location in various relationships and communities; acceptance of the family as the object of treatment and as potentially the most important care-givers (Mudd 13); the sustaining of interaction among patient, family, friends, and hospice staff which entails maximal relief of pain and discomfort compatible with an alert patient; the conviction that death, like life, should occur in communities and not in institutions.

Specific measures implementing these commitments included termination of curative efforts when the best medical judgment indicated they were useless; elimination of all institutional, bureaucratic requirements not absolutely necessary, such as restricted visiting hours, standardized furnishings and attire, and strict structuring of patients' time; elimination of as many distinctons as possible between staff and clients and among staff; freer access to drugs for pain on the assumption that the general hospice setting and freer access would reduce anxiety-fed pain. All of these reflected a conscious shift of focus from the technological attack on the patient's pathology to care for the total person of this still living member of the community.

However, hospice is not antitechnological. It does not reject the very real benefits of medical technology as long as they remain benefits and do not overwhelm other more important considerations. "Technologically specialised curative medicine has catalyzed the growth of hospice medical care as a natural balance," "a care system" which can coexist with the "cure system" (Lack and Fischer 2599). Furthermore, hospice itself employs some important medical technologies, primarily in the area of pain control (Halper 24). But hospice supporters do claim that in the hospice setting such technologies are insistently subordinated to the concern to enhance the lives of persons in community and that they are employed only when their service to that end is unambiguous.

But is hospice a community? Or is it just an alternative mode of health care, or a competing institution? While not a community in the usual sense, hospice exhibits many of the characteristics of authentic community. It emerged in this country not by bureaucratic or governmental fiat but, as in the case of the New Haven hospice, from the people of the community, both in terms of its motivation and its funding (Lack and Fischer), and in response to the demands of patients and their families (Hounder). And as Sandol Stoddard points out, hospice is more often a community-based team than a place.

The clearly value-based nature of both its guiding convictions and its policies and practices places hospice firmly within the category of community as used here. The most persistent and the major unifying theme running throughout hospice is its conviction that truly humane living and dying occur only in the company of supportive, compassionate companions. In fact there is evidence to suggest that hospice, which has "depended on a spirit of compassionate volunteerism" (Brody and Lynn 921) is a sufficiently distinctive health-service community that it attracts a unique sort of nurse who is "extremely genuine, direct, gregarious, spontaneous, capable of warm involvement, and trusting" (Amenta 1984, 418). "[H]ospice works best when the family and the hospice team confront the frightening reality of death together" (Mudd 13). Hospice's aspiration to provide, even to embody such companionship renders unassailable its claim to be a community.

While it is difficult to separate the impact of hospice from other factors, it is clear that hospice has fostered significant positive changes in the technological approach to dying. Perhaps the most specific benefit is "a vastly improved understanding of the physiology and pharmacology of pain control" (Brody and Lynn 921) deriving from hospice's conviction that neither pain nor the means of suppressing pain should be permitted to interfere with the patient's continuing relations with others. Hospice has also alerted those involved in the more traditional treatments to the limitations of what they do (Halper 18, 22; Silverman).

On an ultimately more significant level, Robert Cunningham expresses the hope that "means may be found to turn hospital resources to hospice purposes and use hospice function as a lever to compel changes in staff attitudes and behavior that may be expected to benefit the living as well as the dying" (65). Cunningham's 1979 report that "half the hospices . . . operating or being planned [are] distinct parts of existing hospitals or nursing homes" provides some evidence that this hope is being partially fulfilled. Thus hospice, spawned in its contemporary form by the desire to limit technology's tendency to isolate the dying from community and relationships, by establishing supportive communties to include and accompany the dying, has reaffirmed values which are already getting renewed attention in the very setting hospice found inadequate.

Physicians and the Artificial Heart

In turning to the emergent technology of the artificial heart and to the moral community of the medical profession, specifically physicians, we meet a situation similar in general outline to the two already treated, while again having its own distinctive dimensions. Two of these are of special interest here. The first concerns the complex relations among the community involved, the technology, and the larger society, especially those potentially affected by the technology. Here the same community which may have an obligation to resist and direct the technology is also one of the major communities encouraging and aiding its development, while also standing to gain much from its further development and application. Furthermore, this same community of physicians carries significant responsibility for serving as intermediaries or interpreters between that technology and the rest of society. We have seen that Jehovah's Witnesses bore no responsibility to others for the technology in question, while hospice voluntarily alerted others to an alternative way of dealing with death and dying. Members of this third community, however, simply by occupying the societal role they do, stand as fiduciaries both toward clients and toward the larger society when they evaluate or apply medical technology, including such emerging possibilities as the artificial heart. Any medical technology assessment, resistance, or redirection by physicians is not primarily for their own sake, nor for the sake of their own limited group or community. It is first and foremost owed to the larger society which has enabled them to acquire their skills and knowledge in this area and then has entrusted them with significant powers of self-regulation and peer review for the sake of a healthier society (Camenisch). Thus this community potentially faces a much sharper conflict of interests in its dealings with medical technologies.

The second distinctive dimension of this example is that here we enter a chapter of the community-technology interaction still being played out. Dr. Barney Clark, the first recipient of the artificial heart as a permanent replacement, died on March 23, 1983, 112 days after his operation. Murray Haydon, the third recipient, died on June 19, 1986, 488 days after receiving his artificial heart. William Schroeder died on August 13, 1986, after 620 days of Jarvik support. In addition, the total replacement artificial heart has been used several times as a temporary expedient in the absence of a donor organ (Nieuwsma). In spite of these survival periods, which are impressive for such a boldly experimental procedure, the results achieved thus far by permanent replacement hearts raise the question of whether they can in good consicence and in the name of good medicine be continued at this time. Numerous cerebrovascular accidents and other complications have so compromised the quality of life of the recipients that in December 1985 an FDA advisory committee considered asking DeVries to postpone three other implantations that had previously been approved ("FDA Reconsiders . . ." 16), and even the surgeons who have done the implants have serious disagreements about the justification of the implants at this stage of development (Van 10). Currently, there appears to be an informal moratorium on such replacements.

Moratorium or no moratorium, the question is whether the intervening time is being used to assess this experimental technology. Physicians' involvement and stake in this technology, as well as their specialized professional knowledge, should make them some of society's most competent critics of this technology. But these very characteristics may simultaneously make physicians hesitant to assume that role. The status of all physicians is no doubt enhanced by recent impressive advances in medical technology. Furthermore, it is well known that professionals often mute criticism of their professional peers. And yet both good medical practice with its first principle of *primum non nocere* (first do no harm), and responsible human experimentation with its desire to advance knowledge at minimal cost to experimental subjects, must stand uneasily before the replacement of a functioning human heart—however poorly functioning—with a mechanism which thus far has yielded only highly dubious results.

Certainly, professional assessment of medical technology was a serious issue for the profession and society prior to the emergence of this particular technology. A few months following the death of Barney Clark, Representative Albert Gore, Jr. (D-Tenn.), chairman of the House Science and Technololgy Subcommittee on Oversight and Investigations, summarized the technology assessment situation which does not seem to have changed significantly:

> There are serious shortcomings in the national system of health-care technology assessment. Among these is the fragmentation between the federal government, a variety of professional groups and third-party insurers, all with distinct, overlapping priorities. In this fragmented system, technologies selected for review do not reflect a systematic identification process, but rather a reac-

tion to questions of fiscal reimbursement. . . . [A] vigorous and broadly based health-care technology process must be put in place as soon as possible. Ideally, such a process should include the federal and private sectors . . . and should be separated from reimbursement decisions. (Iglehart 510)

While reimbursement issues, then as now, dominate much medical technology assessment, medical professionals have realized the need for assessment predicated on other criteria. The American Medical Association's Council on Scientific Affairs has for years prepared reports for the profession on especially popular or controversial medical technologies. In 1982 the AMA's Diagnostic and Therapeutic Technology Assessment (DATTA) program was inaugurated to provide physicians with brief, prompt answers to questions about the efficacy and safety of various technologies (Jones 387–88). And one of the chief functions of the major medical journals is, through the publication of both experimental findings and clinical experiences, to encourage and facilitate the assessment of both established and new diagnostic and therapeutic technologies.

Is such technology assessment, largely confined to professional circles and media, adequate to meet the larger public's legitimate interests? Well-publicized technologies arouse general interest, raise the hopes of potential beneficiaries, and introduce a wide range of social issues such as funding priorities, distribution and access, and reimbursement. Should the public have to wait for the several months or more usually required for journal publication to know what competent professionals think of the Baby Fae baboon heart replacement, or whether adequate experimental groundwork had been laid for the artificial heart?

At the very least we must note that there has been no massive public outcry by physicians concerned about the precipitous (if such it was) application of such technologies to human subjects. One physician's private reference to the Jarvik artificial heart as the "Jarvik stroke machine" has no doubt been duplicated by others, and Dr. Stuart Frank, chief of cardiology at Southern Illinois University School of Medicine, Springfield, did venture in print that he would not want his child to go through what Baby Fae suffered (Gorner and Spencer). Even the surgeons who have performed the various implants have aired some of their disagreements about the benefits of the current technology (Van 10). But such guarded general comments comprise at best a piecemeal response inadequate to guide public opinion and policy.

Other more focused, and thus potentially more helpful and informative, comments on these procedures have concerned the adequacy and clarity of the consent forms used (Annas 1983; Cassell 32); the

126

insufficient attention given to what happens when a replacement heart with high output is placed in a body which has adapted to an organ with low output (Cassell 27); the "restricted, dreary existence" which seems to be the best these patients can hope for (Gorner and Spencer); our inability to know how long the artificial heart recipients would have lived had their own hearts not been removed; the confusion between therapy and research which characterizes some of the thinking about these procedures; whether at least one of the artificial heart recipients was as close to death as had been claimed (Preston 5); the commercial interests in the artificial heart of some of the University of Utah medical staff (Reiser 172); how the development of the artificial heart technology will affect all patients; and how it might further distort priorities in the expenditure of research funds (Preston 5, 7). Telling and troubling though some of this criticism is, it is inadequate to inform the public if we are flirting with a repetition of "one of the darkest episodes in the annals of surgery"—the "unseemly rush" to 150 heart transplants in the two years following Christian Barnard's pioneering transplant (Annas 1985, 15).

In addition, Humana Hospital of Louisville has been sharply criticised for how it publicized its first venture into the artificial heart replacement arena with William Schroeder. This campaign included well-orchestrated news conferences and releases, massive get-well cards suitable for feature news pictures, and, according to Milton Nieuwsma, even distortion of the facts of both the history, and the current situation of the permanent heart implant. This episode inevitably raises the question of the possible influence of the profit motive on this already complex situation.

The above professional responses to the artificial heart replacement and related interventions are negligible in neither variety nor content, given the complexity and controversial nature of this area, as well as the number of physicians and physician agencies directly involved or potentially affected. Yet there is strikingly little public comment on these matters, and even less systematic assessment available to the public. Several considerations help explain this fact: the concern that too much public discussion might discourage potentially beneficial investigations (Jones 388); the tendency of some physicians to rest their assessment of such procedures on the adequacy of the informed consent obtained (Gorner and Spencer), an approach which may inappropriately mute assessment of the procedure itself; the argument that physicians not directly involved in such procedures must at least be very circumspect critics since they cannot understand the agony and the pressures of waiting for donor organs while life ebbs away (Breo 3). While such assessment-retarding factors, according to Thomas A.

Preston, M.D., "combine to favor the independent course of Dr. De-
Vries, his medical team, and Humana, Inc. . . . [t]ime and a rising in-
sistence for public accountability are working against them" (5).

One form of assessment-oversight that should be noted is the Institutional Review Board (IRB), which can be assumed to be operative
in any institution where such procedures are now performed. But
such boards, whose findings are generally not made public and who
function within particular institutions, cannot make up for any lack
of more comprehensive professional assessment in the larger public
forum.

Conclusions about the professional medical assessment of the artificial heart drawn from the above evidence will vary widely. Some
will argue that aside from a few of the rasher critics, what we currently have from physicians and researchers is precisely what the
present situation calls for—a cautious wait-and-see approach adopted
by professionals who are aware of the numerous complexities involved. Other less sanguine observers will see here an unfortunate absence of professional courage which protects professional interests
and peers' reputations while generating in actual and potential pa-
tients-subjects and their families unsustainable hopes and exposing
them to procedures which are not only unproductive but extraor-
dinarily burdensome. Only time and future developments, including
more prompt, accessible, and coordinated information from knowl-
edgeable professionals, will permit laypersons to answer the two
complex questions facing them in this matter: Is the artificial heart
ready for human use, even at the experimental level, and have physicians been as helpful as they could and should have been in assuring
that the public received helpful data about the artificial heart, its potential risks and benefits?

Conclusion

Two sharply focused success stories and one unfinished chapter do
not provide a conclusive answer on our ability to direct technology to
the service of broader and deeper human interests and goals. Nevertheless, several limited conclusions are justified. First, neither the
technological imperative nor the momentum of emerging technologies need overwhelm all countervailing values. Second, specific technologies can be resisted or redirected in their applications to limited
communities. Third, those efforts may positively affect the use of
those technologies in the larger society. Fourth, one of the major conditions for such resistance and redirection is the existence of a community which perceives in the technology significant disvalue, and

whose shared values provide it with a place to stand as it tries to limit or redirect technology. Jehovah's Witnesses and the hospice movement clearly possess such a set of values. Currently it is not clear that contemporary medicine retains the unifying set of values which once made it a well-focused and thoroughly moral undertaking. If it has lost its firm attachment to those values, or if they have become so diverse and diffuse that they cannot ground a coherent response to advances in biomedical technology, then we are left with the troubling questions of how and by whom such advances will be assessed and directed.

Works Cited

AMENTA, MADALON O'RAWE
1984   "Traits of Hospice Nurses Compared with those who Work in Traditional Settings," *Journal of Clinical Psychology* 40: 414–20.
1985   "Hospice in the United States: Multiple Models and Varied Programs," *Nursing Clinics of North America* 20: 269–79.

ANNAS, GEORGE J.
1983   "Consent to the Artificial Heart: The Lion and the Crocodiles," *Hastings Center Report* 13, no. 2: 20–22.
1985   "At Law: The Phoenix Heart: What We Have to Lose," *Hastings Center Report* 15, no. 3: 15–16.

BONAKDAR, MOSTAFA I., ARTHUR W. ECKHOUS, BURTON J. BACHER, ROMEO H. TABBILOS, and DAVID B. PEISNER
1982   "Major Gynecologic and Obstetric Surgery in Jehovah's Witnesses," *Obstetrics and Gynecology* 60: 587–90.

BREO, DENNIS L.
1984   "Why Baby Fae's Doctor Will Dare to Try Again," Chicago *Tribune* (Nov. 18): section 2, 1–3.

BREWSTER, D.C., J. J. AMBROSINO, R. C. DARLINS, J. K. DAVISON, D. F. WARNOCK, A. R. MAY, and W. M. ABBOTT
1979   "Intraoperative Autotransfusion in Major Vascular Surgery," *American Journal of Surgery* 137: 507–13.

BRODY, HOWARD, AND JOANNE LYNN
1984   "Sounding Board: The Physician's Responsibility under the New Medicare Reimbursement for Hospice Care," *New England Journal of Medicine* 310: 920–22.

CAMENISCH, PAUL F.
1983   *Grounding Professional Ethics in a Pluralistic Society*. New York: Haven Publications.

CASSELL, ERIC J.
1984   "How is the Death of Barney Clark to Be Understood?" In *After Barney Clark*: 25–41. Edited by Margery W. Shaw. Austin: University of Texas Press.

CLARKE, J. M. F.
1982   "Surgery in Jehovah's Witnesses," *British Journal of Hospital Medicine* 27: 497–500.

CORBETT, TERRY L. and DOROTHY M. HAI.
1979   "Searching for Euthanatos: The Hospice Alternative," *Hospital Progress* 60 no. 3: 38–41, 76.

CUNNINGHAM, ROBERT M., JR.
1979   "When Enough is Enough," *Hospitals* 53: 63–65.

DIXON, J. LOWELL, and M. GENE SMALLEY
1981   "Jehovah's Witnesses: The Surgical/Ethical Challenge," *Journal of the American Medical Association* 246: 2471–72.

DORNETTE, WILLIAM H. L.
1973   "Jehovah's Witnesses and Blood Transfusions: The Horns of a Dilemma," *Anesthesia and Analgesia* 52: 272–78.

_____
1985   "FDA Reconsiders Artificial Heart," Chicago *Tribune* (Dec. 3): section 1, 16.

FARR, A.D.
1972   *God, Blood, and Society*. Aberdeen: Impulse Books.

FINDLEY, L. J., and P. M. REDSTONE
1982   "Blood Transfusion in Adult Jehovah's Witnesses: A Case Study of One Congregation," *Archives of Internal Medicine* 142: 606–67.

_____
1952   "Foreign Letters: Denmark," *Journal of the American Medical Association* 148: 953–54.

GONZALEZ, ELIZABETH RASCHE
1980   "Medical News: Fluosol a Special Boon to Jehovah's Witnesses," *Journal of the American Medical Association* 243: 720–21.

GORNER, PETER, and JIM SPENCER
1984   "Peers Agree with Surgeon's Decison," Chicago *Tribune* (Nov. 18): section 2, 3.

HALPER, THOMAS
1979   "On Death, Dying, Terminality: Today, Yesterday, and To-morrow," *Journal of Health Politics, Policy and Law* 4: 11–29.

HARRIS, T. J., N. R. PARIKH, Y. K. RAO and R. H. OLIVER
1983   "Exsanguination in a Jehovah's Witness," *Anesthesia* 38: 989–92.

HOUNDER, SUSAN W.
1979   "Letters: The Hospice Movement," *Journal of the American Medical Association* 241: 2600.

IGLEHART, JOHN K.
1983   "Health Policy Report: Another Chance for Technology Assessment," *New England Journal of Medicine* 309: 509–12.

JONES, RICHARD J.
1983   "Editorial: The American Medical Association's Diagnostic

and Therapeutic Technology Assessment Program," *Journal of the American Medical Association* 250: 387–88.

Krant, Marvin J.
1981   "Hospice Philosophy in Late-Stage Cancer Care," *Journal of the American Medical Association* 245: 1061–62.

Lack, Sylvia A. and William Fischer
1979   "Letters: The Hospice Movement," *Journal of the American Medical Association* 241: 2599–2600.

Mudd, Peter
1982   "High Ideals and Hard Cases: The Evolution of a Hospice," *Hastings Center Report* 12, no. 2: 11–14.

Munley, Anne
1983   *The Hospice Alternative: A New Context for Death and Dying.* New York: Basic Books.

Nieuwsma, Milton
1984   "Hyping the Artificial Heart," Chicago *Tribune* (Dec. 7): section 1, 23.

Ott, David A., and Denton A. Cooley
1977   "Cardiovascular Surgery in Jehovah's Witnesses," *Journal of the American Medical Association* 238: 1256–58.

Preston, Thomas A.
1985   "Who Benefits from the Artificial Heart?" *Hastings Center Report* 15, no. 1: 5–7.

Reiser, Stanley J.
1984   "The Machine as Means and End: The Clinical Introduction of the Artificial Heart." In *After Barney Clark:* 168–75. Edited by Margery W. Shaw. Austin: University of Texas Press.

Saunders, Cicely
1977   "Dying They Live: St. Christopher's Hospice." In *New Meanings of Death:* 153–81. Edited by Herman Feifel. New York: McGraw-Hill.

Schaller, R. T., Jr., J. Schaller, A. Morgan, and E. B. Furman
1983   "Hemodilution Anesthesia: A Valuable Aid to Major Cancer Surgery in Children," *American Journal of Surgery* 146: 79–84.

Silverman, William A.
1982   "Commentaries: A Hospice Setting for Humane Neonatal Death," *Pediatrics* 69: 239.

Smith, David H., and Judith A. Granbois
1982   "The American Way of Hospice," *Hastings Center Report* 12, no. 2: 8–10.

Smith, Earl Belle
1980   "General Surgery in Jehovah's Witnesses—Personal Experience: A 22-Year Analysis," *Journal of the National Medical Association* 72: 657–60.

Stoddard, Sandol
1980   "Letters: The Hospice Challenge," *New England Journal of Medicine* 302: 1314.

PAUL F. CAMENISCH

TAYLOR, MICHAEL
    1982   *Community, Anarchy and Liberty*. New York: Cambridge University Press.

VAN, JON
    1985   "Surgeons Debate Value, Future of Artificial Hearts," Chicago *Tribune* (Oct. 27): section 1, 10.

9

Community, Time, and the Technical Order

ROBERT ROTENBERG

The promotion of scientific and technological development should be carried on for the benefit of employees and in the interest of the entire economy; the increase in productivity due to automation is to provide material, social, and humanitarian improvements for the employees.

Hertha Firnberg, Austrian Minister for Science and Research

Introduction: Technology and Time

To live in a community is to share a structure of activities. The position of work activities, mealtimes, family life, and sleep in the daily round is determined neither by nature nor by individual households. This structure of time is rather the product of community life.[1] With it

1. Communities create schedules by making people publicly aware of the temporal commitments created by their various institutions. Each activity within a schedule has four different dimensions: a *duration*, a position in the *sequence* of the daily or weekly round, a relation to simultaneously occurring activities (*timing*), and a rate of repetition (*tempo*). In this discussion, I use the word *rhythm* to refer to one or more of these dimensions.

Relationships between people and institutions demand different time commitments or *time demands*. Some of these are closely tied to the rhythm of machines, such as work time demands; others, such as parenting, less so. These time demands accumulate, eventually occupying nearly all of the available time not given over to sleep and personal needs. These recurring demands produce the sense of routine in our lives.

we predict fairly accurately where and when to contact another community member. Because of it we feel left out of things when odd work shifts or family responsibilities draw us away from the rhythm of the majority. The most powerful determinants of any community's structure of activities are the demands made on us by our systems of production and reproduction, work and household. The relevance of a discussion of community time structures in a collection of papers on the cultural impact of new technologies is precisely this: New technologies change the way people work by altering the organization of work activities in the daily round. Such changes in work life directly affect the time available for household and individual activities and do so in surprising and unpredictable ways.

New technologies invite contrasting visions of how life will change after we become accustomed to using them. One vision sees only good, while the other sees only the fires of hell. Both sides impute to technology a power to transform society which it does not have. Technologies are only used to achieve ends set by and for people. The impact of new technologies on community should be attributed to the decisions made by the people who control the use of those technologies to achieve their desired ends. I call these ends, as well as decisions made to realize them, the *technical order*. The community of households enters this decision-making process because all people, and not just those working with the new technologies, must find solutions to myriad problems which these narrowly focused activities inevitably create.

Two examples from the European press from 1984 provide examples of how communities are affected by technological change. *Le Monde*, the liberal Parisian newspaper, envisions two neighbors, both employed by the same insurance company, working at home, sharing the same job through a computer link-up, and each working half a day. Their work day is *half* as long as it once was. But each ends up potentially working much longer. The French Ministry of Industry and Research predicts that people will not know what to do with the newly won time and will seek out multiple jobs to fill their time, spending "ten hours a day at two jobs instead of eight hours at one" (Herteaux 1984).

The issue, as the French see it, is not whether to reduce work time, but how best to do so. They don't believe household activities and recreation are so elastic that they can suddenly and meaningfully expand to fill four hours of time. The purported four-hour work shift challenges the household to find or invent recurring activities where none previously existed. These work lives are the product of decisions made by the managers of a hypothetical insurance company, managers who enable their employees to work at home, to share a job, and to work for shorter periods on a regular basis. Although these decisions originate as rational strategies on the part of managers, the workers and their families must then deal with the problems generated by any changes in their routine.

Or take this report from *Wirtschafts Woche*, a business weekly from Germany, on BMW's new, highly automated plant at Regensburg. In this plant, rather than replacing the work force, the robotic assembly line actually requires a *more* highly skilled work force because "the technology promised much more than it could deliver." This shortfall in expectations creates the demand for a work force which can watch for consistency in materials, one with the independence to determine when to shut down the line to make adjustments. In fact, the more automated and expensive the production process, the more important the human factor in maintaining it. Both BMW management and the automotive trade unions see a movement away from the "skills polarization" pattern, which has segregated design and programming from the actual operations on the production floor, and toward a more highly trained and flexible work force which can oversee maintenance and repair, as well as production, and which can program the tools for new jobs. At the present time, the report continues, both the old and the new patterns exist in German factories ("The Human Factor"). At some future point production will evolve to utilize these new workers more effectively. But at present and for the immediate future, they remain overqualified machine tenders.

The monotony such highly skilled individuals must endure in this transitional production line, when their skills are only demanded in the most rare and extreme situations, poses a problem for the community. The work life depicted in this report is akin to that of air traffic controllers, who must either devote all of their attention to even the most routine of tasks or court disaster. Short shifts, high pay, and high professional status in the community compensate these workers for their work stress. What is the effect on the nonwork life of people and their households as more and more workers are asked to take on such stressful work lives? What are the needs of such workers when their shift ends? Can the existing household and leisure institutions meet the challenge of these workers, returning them to the workplace the next morning refreshed and ready to engage in this productive activity again? As with the French insurance company, community-based problems are engendered by the technical order which chooses and implements the new technology. The problem lies in the relationship between workers on the robotic assembly line with their communities, and not in the robotic technology itself.

135

Why do these German and French observers seem more conscious of the costs and benefits of this new technology. Do the technical orders of Europe and the United States differ sharply? These are not idle questions. The best protection against such problems is to understand how the technical order is constituted. Only with an awareness of the values of the people who control technology can communities predict how the change will affect them and act accordingly. Since cooperation between work institutions and the community reduces the social costs of technological change, these Europeans display a consciousness which may give them an advantage in planning.

To show how the values of the technical order evolved in American society, this essay samples the historical responses of communities in the United States to secondary industrialization in the later half of the nineteenth century. In an effort to understand why the European observers might be more conscious of the costs and benefits of new technologies, I also offer observations from my research in Vienna, Austria, in the 1970s and 80s, a postindustrial technical order which exhibits contrasting attitudes toward the deployment of technological activities in time.

Managing Time through Technology

Throughout the period from 1815 to 1920 the United States engaged in a process of industrialization from which it emerged as a major inventor, producer and user of complex technology. This process took the particular direction of increasing dependence on substituting technologically based manufacturing for human craftsmanship in order to lower cost. Why this occurred is in large part traceable to the values of the technical order current among industrial managers, especially during the period of 1848 to 1914. The process of industrialization during this period created the role of manager and in so doing transformed the technical order. The cost-effective meshing of worker and machine became both the content and the *raison d'être* of the role of manager. The particular goals of the technical order of the first managers already included a high priority for research and development aimed at making the production process more efficient.

The quest for efficiency did not become a dominant value in this technical order because it is an objective quality toward which rational systems always develop. Rather, it became a dominant value because of the failure of most U.S. manufacturing centers to establish a stable, disciplined, local work force. Throughout this period, as Herbert Gutman has convincingly documented, U.S. work institutions depended on migratory labor from premodern communities in Europe, Latin

America, and East Asia. Of course, the cost of labor would have been substantially higher if the same work institutions were forced to meet the rising expectations of the initial industrial work force of the 1815 to 1848 period. The children of the work force were spatially and socially mobile because of social and geographical features unique to the American experience. They easily moved on to nonfactory occupations in the expanding industrial economy, and were replaced in their parents' factories by new migrants. The work force of the factory system was constantly repopulated by people unsocialized to the discipline of factory life, but also comparatively undemanding in their wage expectations.[2] The task of ensuring effective use of time and machines fell to the managers.

Industrial time discipline differs sharply in its organization of activities from any other kind of work in human society. Pre-industrial craftsmen completely controlled the pace of their work activities. Workers did whatever was necessary to complete a contract by a given date. When ahead of schedule, work pace could slacken; when behind, it would accelerate.

The nineteenth century factory system in both the United States. and Europe was predicated on wresting this control from undisciplined workers and first putting it in the hands of the factory owner and, later, the manager. Tool and worker became equivalent factors in the manager's effort to reduce production costs. The time demands upon the worker came from a manager who commanded work on a machine for a specific period of time. It was the authority of the manager that determined the pace and not the work itself.

If the disciplined work habits of the factory had to be forced upon the initial generation of workers, subsequent generations of workers were raised entirely within the expectations of the new technical order. This second generation resigned themselves to being managed with less cost to the factory. However, in a peculiarly American development, the high degree of mobility of these workers removed any long-term savings from this effort. As soon as one generation of workers was socialized to conform to the rigid durations and sequences of factory life, it would move up and be replaced by new immigrant workers from rural areas with pre-industrial work habits. The projected savings from hiring the children of disciplined workers who could accommodate the demands of factory life with less direct super-

2. As exceptions which prove the rule, older New England industrial towns, where one large factory was the sole wage institution, could often capture a worker community for a number of generations. At the Amoskeag Manufacturing Company of Manchester, New Hampshire, generations of workers carried their community ties into the factory, ultimately controlling the hiring of their children as new workers (Hareven).

vision were negated. In each succeeding generation, American factory managers found themselves faced with the task of socializing a new group of workers to the discipline of their technical order.

Early factory managers devised two strategies to handle this recurring turnover in experienced workers: piece-rate organization and technology substitution. By tying workers' wages to a specific production quota based on the pieces actually completed, managers could more effectively control the productivity of individual workers. This was especially important during the initial, settling-in period, when the work discipline would appear most alien to new workers.[3] This strategy appealed to labor-intensive industries. Once immersed in the factory system through piece-rate organization, the new workers were more likely to accept the time-rates of more capital-intensive enterprises.

As the workers gained experience, they stepped up to better pay and to greater involvement in technologically assisted work. In these factories, worker discipline and compliance with the temporal structure of factory life became a more important factor in setting production costs. According to Daniel Rodgers, managers hit upon the technique of technology substitution quite early. They invested in developing "self-acting" machines and processes which can substitute for human labor at the most intricate and costly points in the process, thereby greatly reducing the work force to the smallest feasible number of skilled machinists (66).

Clearly, this technology could have been implemented in a variety of ways. From the workers' point of view, technology substitution held the potential for freeing them from drudgery in the production processes, allowing them to devote their efforts to the more skilled jobs of repairing and maintaining their mechanical liberators. From the managers' point of view, technology substitution held the promise of removing workers from those pressure points in the productive process where the required skills or costs of mistakes were the greatest. The historical drift of manufacturing in the United States supported those inventions which could replace workers, redistributing the jobs within the factory so that the humans performed the dullest, most repetitive tasks, and machines the most interesting ones.

How does this technology substitution affect workers' time de-

3. Piecerates mimic the older pattern of work organization by tying work to the completion of a specific product. They differ from the older tradition by insidiously dividing the fully completed product into a large number of semicomplete parts, none of which could by itself give the worker a sense of work completed, and by setting minimum rates so high that the full shift is necessary, at least initially, for the worker to achieve it.

mands before, during, and after work? The chief consequence lies in a separation in the minds of workers and managers between one's time demands at work and the time demands of one's household. Decisions about time commitments at work happen as if the other commitments did not exist. Thus, opportunities for overtime, shift changes, and more recently, flexitime, the four-day workweek, and telecommuting,[4] are offered to workers only when it is both efficient and cost-effective for the firm to do so, and withdrawn when such measures fail to meet managers' expectations. Managers may say that an innovative work pattern is offered for the convenience of workers, but subsequent inconvenience is quickly forgotten when a program is scrapped. The high cost of equipment, rapid changes in the technology, and a reluctance on the part of managers to let loose the reins are currently stopping more industries from experimenting with telecommuting. It is evident that the values of the nineteenth century technical order continue to define the management of technological change in the United States today.

The lives of community members also are shaped by time demands originating in the nineteenth-century technical order. For example, the retail markets servicing households in American society are forced to schedule the shifts of their workers to meet busy periods generated by the time demands of the household in a market's area. With more multiple-wage-earning families, fewer household members are available to shop during the day. This results in an increased number of retail clerks forced to work evening and weekend shifts, which in turn changes the periods of time these workers can spend in their homes. Any large-scale experiments with telecommuting would make more shoppers available during ordinary day shifts and somewhat reverse the trend. Because people cannot be two places at once, when all work institutions manage the time of their employees in the same fashion, the result is tidal movements of people to and from work, and to and from markets. But in all cases, it is the schedules of households, markets, schools, and public services that must react to

4. The four-day week arose in the oil crisis of the 1970s as an energy efficient alternative to the standard five day, 40-hour week. At its height in 1975, a mere 2.2 percent of the work force were involved in this experiment. Flexitime is the system of allowing employees to set their own work hours. This strategy is prevalent among service and transport personnel (11%) and in middle-level management positions (2.5%). It enables commuting to take place at off-peak times of the day (Tefft 11). Telecommuting is the name given to the computer-linked office at home. As of early 1986, only 30,000 workers were working full-time as telecommuters, while 100,000 were working part-time. The major component of this work force are mothers who are trying to combine work with child care, and who intend to return to an office environment when their children go to school (Noble 22).

the needs of the work institutions. The technical order produced by the U.S. experience is so pervasive, we cannot conceive of it being otherwise.

A Case of It Being Otherwise

My research into the organization of household and public schedules in Vienna, Austria, shows the degree to which this technical order is related to the global phenomenon of capitalist industrialism and the extent to which it is specific to patterns of worker socialization in Austria in the nineteenth century (Rotenberg). The Austrian experience with the industrialization of work bears both similarities and differences to that of the U.S. The major differences lie in the development of a very strong trade union movement and the effective political control of public policy by this movement through the various Socialist parties that have ruled the Vienna since the 1900s and Austria since the 1960s. While in control of these governments, the Socialists generated policy that specifically supported public control over work time.

The contrasts between Austria and the U.S. lie in the differing historical development of public policy with respect to the rhythm of work. Central European industrialization, in which the Austrian experience was embedded, extends to the mid-eighteenth century. This region also experienced a large migration of workers to the growing metropolitan industrial centers during the mid-nineteenth century. But most of these migrants were already disciplined industrial workers. They migrated along with their factory jobs from the smaller regional towns to the important marketing and transportation centers. This was a horizontal movement only. Mobility was very difficult in this highly class-conscious empire. For this reason, more children were available to replace their parents in the same factories. When these factories moved into places like the suburbs of Vienna, they brought their factory-disciplined families with them. Thus, the costs of work socialization in Austria were, over the long run, far less than they were in the United States in the same period (Matis; Meissl).

Nineteenth-century factory conditions were quite similar to those in the United States. Both piece-rate and technology substitution strategies were employed. As in the United States, technology substitution relegated workers to the least interesting parts of the production process. Shifts began before sunrise and ended long after sunset.[5] In such atrocious working conditions the workers' trade union

5. These work days were so long in the period of 1830s and 40s that the marriage rate fell among the immigrant worker communities. The time demands of work roles were

movement grew interested in using technology substitution for the benefit of workers. As early as the 1890s, Socialists demanded that all technology substitutions should result in greater freedom from drudgery for workers, not in the removal of workers from sensitive positions in the factory process. These less mobile communities become dimly aware that control over the rhythm of the work day was a more important battle than wage gains. As the quote from Minister Firnberg which opened this essay demonstrates, the position of the Socialist party remains unchanged.

Except for the mobility differences in the two work forces, the industrialization experience was and remains quite similar. The managers of Austrian industry study the same textbooks, deal with the same day to day problems, read the same trade journals, and belong to the same professional associations as American managers. Indeed, they compete with each other. But differences in the consciousness of its workers transformed Austrian society in ways which limited the manager's control over work time.

Although the principle of public control of the length of the work day is accepted in most industrialized countries, the Austrian trade unions took almost seventy years to put this rule of law into effect. But unlike the American experience with the eight-hour day or the forty-hour week movement, the politically powerful Austrian trade union movement was able to *enforce* the principle of public control over the work rhythm in the retail, private, and the other nonunionized parts of the economy. Each step in this widening process of public control over the schedules of public institutions represented a constraint on the options of both work institutions and households. These included the reduction of the work week, the setting of minimum vacation days and holidays, the creation of store and warehouse closing ordinances, or the maintenance of the two-hundred-year-old tradition of a six-day school week.[6]

---

so great that all nonwork time demands, including the formation of households and their maintenance, became less and less tenable (Ehmer 129).

6. The first public control of work time legislation, a ten-hour day, was won in 1891 but not effectively enforced. The eight-hour day ordinance was first passed in Austria in 1919, immediately after the break-up of the Austro-Hungarian Empire. A 48-hour week limit on regularly paid work was passed in 1948. In 1959, the standard workweek was reduced to 45 hours, and in 1975 to 40 hours (Zeisel). At present, a number of individual trade unions, notably the printers' union, have individual contracts for as few as 32 hours a week. The Socialist party has set the goal of a national reduction to a 35-hour standard within the next decade.

The government increased the number of national holidays to thirteen in 1946. By 1980, the minimum number of paid vacation days was twenty-five. With the right combination of movable holiday dates close to the beginning or end of weekends, an ordinary factory worker can end up working less than 170 days a year.

For example, the 1958 store closing ordinance requires all retail stores except food shops to close each weekday at 6:00 P.M. This is true of both family-run shops in the neighborhoods and department stores in the central business district. Regardless of whether customers are present, a retail manager cannot ask employees to work later than 6:00 P.M. By setting a store closing time, the government ensures retail clerks time with their families without fear of pressure from employers.

This ordinance has met stiff resistance from households with two wage earners who demand access to shops in the evening. These demands are thrown back at employers by the government. They argue that employers create these problems when they place their work shift changes too close to the store closing time. Thus, it is not the concern of the government, whose role is to protect the nonwork time of employees of small firms from overtime exploitation. The fact that this Austrian policy receives political support speaks to workers' consciousness of their employers' technical order. Having emerged from the experience of previous generations, this consciousness lends support to policies that restrict work time for the benefit of households rather than the marginal utility of the firm.

The idea that work institutions should conform to the time demands of members of households is alien to the technical orders of both Austrian and Americans managers. The difference is that in Austria, higher consciousness of the need to protect the home life of workers is effective in shaping the rhythm of work; in the United States, it is not.

Conclusion: Community, Time, and Contemporary High Tech

This essay began with an assessment of the role of communities during periods of technological change. We have seen that communities share problems generated by managerial decisions over the implementation of new technologies. These problems arise precisely because the schedules of community members include work activities, but the schedules of work institutions do not include community activities. Differences in the development of the worker communities between U.S. and Central European industrial histories point out that the technical orders of managers originated in the problems of worker discipline early in the industrial experience. This experience is also responsible for the differing levels of consciousness in worker communities over the need to protect their home lives from managerial decisions ostensibly directed to their work lives.

Management of technology, I wish to repeat, is not "natural" but historically derived. This puts the responsibility for the effects of technological change squarely on the decisionmakers. Nonimpoverishing, nondisruptive alternative models are available. The humanization of work must begin with an enhanced picture of the worker's life outside the work place. This means placing the needs of households on a par with the needs of the technical order. The Austrian model may not be an effective one for the United States. Our consciousness problems and solutions in the workplace as more appropriately originating outside of government. The Austrian case nevertheless demonstrates that an alternative consciousness and alternative solutions are possible.

Works Cited

EHMER, JOSEF
1982   *Familienstruktur und Arbeitsorganization im frühindustriellen Wien*. Wien: Verlag für Geschichte und Politik.

FIRNBERG, HERTHA
1974   *Welcoming Remarks*. Conference on Human Choice and Computers, International Federation for Information Processing. Vienna: Union of Clerical and Technical Employees in Private Industry.

GUTMAN, HERBERT
1976   *Work, Culture and Society in Industrializing America, 1815–1919*. New York: Knopf

HAREVEN, TAMARA
1982   *Family Time and Industrial Time*. Cambridge: Cambridge University Press.

HERTEAUX, MICHEL
1984   "Taking Work Home: A Comeback for Cottage Industry," *Le Monde* (Sept. 16). Reprinted in translation in *World Press Review* 32 (Jan. 1985): 38.

————
1984   "The Human Factor: The Indispensable Nonrobotic Worker," *Wirtschafts Woche* (October 19). Reprinted in translation in *World Press Review* 32 (Jan. 1985): 39.

MATIS, HERBERT
1968   "Die Ansätze zur Industrialisierung in Niederösterreich im Spiegel einer zeitgenössischen Reisebeschreibung," In *Tradition, Zeitschrift für Firmengeschichte und Unternehmerbiographie* 13: 119–132.

MEISSL, GERHARD
1980   "Industrie und Gewerbe in Wien 1835 bis 1845. Brachenmässige und regionale Strukturen und Entwicklungstendenzen im Spiegel der Gewerbeausstellungen von 1835, 1839, and 1845." In Banik-Schweitzer et al., *Wien im Vormärz*: 75–106. Wien: Verein für Geschichte der Stadt Wien.

NOBLE, KENNETH B.
1986   "Commuting by Computer Remains Largely in the Future,"
            New York *Times* (May 11): Section 4, 22.
ROGERS, DANIEL T.
1978   *The Work Ethic in Industrial America, 1850–1920.* Chicago:
            University of Chicago Press.
ROTENBERG, ROBERT
1981   "The Impact of Industrialization on Mealtimes in Vienna,
            Austria," *Ecology of Food and Nutrition* 11:25–35.
TEFFT, SHEILA
1981   "4-Day Workweek Doesn't Work If Firm's Business 'Out of
            Sync,'" Chicago *Tribune* (Nov. 9): 11.
ZEISEL, PETER
1971   *Die Geschichte der Arbeitzeitregelungen.* Master's Thesis. Hoch-
            schule der Welthandel, Wien.

10

Community Planning for Technological Development
A New Bargaining Process

MARC A. WEISS

AND

JOHN T. METZGER

This article outlines a process of "community collective bargaining" where community representatives join workers, corporate managers, and investors in negotiating trade-offs to achieve conflicting goals for technological development. The bargaining model is constructed from experiences in Chicago and Pittsburgh. The Chicago case, in which neighborhood organizations bargained with the city's large banks to increase inner-city lending, provides a general framework for understanding broadly based negotiations over structural economic issues. The Pittsburgh case offers an example of this negotiating framework applied to high-technology development.

High-technology industries are increasingly becoming the focus of attention for economic planners around the country. State and local governments are designing programs to stimulate technological innovation and the growth of high-technology industries. These economic development programs, geared to attracting and supporting new and expanding high-technology firms, range from policy development, education and training to support for research, technical and management assistance, and financial assistance (Peltz and Weiss). While many of these initiatives are new and difficult to evaluate, their design and implementation often suffer from a lack of balanced representation among the various constituencies with a stake in community planning for high-technology industries. Public sector policymakers who seek to create a consensus over goals for high technology development are often constrained by the imbalance of power among businesses, workers, citizens, and other political and economic interests.

States and cities that have tried to avoid the traditional approach to economic development have been stymied. In California, the administration of Governor Edmund G. Brown, Jr., attempted to do this by bringing together corporations, entrepreneurs, labor unions, and citizen groups through the California Commission on Industrial Innovation (CCII). The existence of CCII created an arena for public-private bargaining or "quid pro quos" within the high-technology sector (CCII; Weiss 1984).

The Brown administration was unable to achieve the desired outcomes of balanced representation and political consensus. The possibility of consensus was hampered by the state's inability either to control the necessary resources or to mobilize a broad-based coalition of support enough to influence the development of high-technology industries. As a result, the state's high-technology development efforts became fragmented, consisting of a "string" of initiatives linked more by immediate concerns than by effective long-term strategies (Weiss 1984).

The California experience indicates the limits faced by government in addressing the impact of high technology development. In the absence of an informed and mobilized political constituency of citizen and community interests, policymakers lack the power and resources to broker effectively with business interests over the direction and outcomes of economic growth. This conclusion points to another possible policy approach, one in which government uses its regulatory and spending powers to convene a direct, collective bargaining process between community coalitions and private corporations over economic development policy.

This strategy has been outlined in the direct negotiations between a coalition of neighborhood organizations in Chicago and the city's largest financial institutions over inner-city real estate lending practices. The community groups have acted as spokesmen for real estate developers, sitting at the bargaining table with real estate lenders to negotiate a plan for privately financing community redevelopment of inner-city neighborhoods.

The strategy of negotiated financing for neighborhood revitalization is made possible by a federal law called the Community Reinvestment Act (CRA). The CRA essentially requires legally chartered financial depository institutions to loan money in the geographic areas from which they draw their deposits. The CRA and its companion legislation, the Home Mortgage Disclosure Act (HMDA), were passed by Congress in the mid-1970s as a response to the "redlining" issue, in which various community groups documented patterns of disinvestment by lending institutions in certain central-city neighborhoods. The HMDA required public disclosure of lending patterns so that the performance of banks and savings institutions can be evaluated by community groups and public officials. Citizens can obtain access to HMDA data and utilize it to challenge an institutional lender's CRA record. A CRA challenge can cause federal regulators to deny permission for lenders to add new branches or to merge with or acquire other institutions.

These two tools have been used by the Chicago Reinvestment Alliance, a broad coalition of community organizations, neighborhood development groups, and city-wide development networks from across Chicago who share common experiences in both facilitating neighborhood reinvestment and developing employment and housing opportunities. In the spring of 1984 the Alliance successfully negotiated neighborhood lending agreements with three of Chicago's four largest banks which were seeking approval from federal regulators for their expansion plans. First National, Harris, and the Northern Trust Banks agreed to create an aggregate, five-year lending pool of $173 million for housing, commercial and industrial development in inner-city Chicago neighborhoods. The Alliance has also negotiated a $20 million agreement for home improvement loans with Continental Illinois Bank, another large Chicago institution that was seeking to expand its operations. The Chicago case is perhaps the most significant community-bank partnership of its kind in the country, due to the size and scope of the agreements, as well as the role played by community-based organizations in loan packaging and program review.

The most distinctive feature of this model of negotiated development is that it is based primarily on private initiative—no government agency formulates or implements a program for neighborhood reinvestment. Instead, strategies are both created and managed by community organizations in tandem with financial institutions. The government acts (a) as a formal rule maker, setting the context for negotiations through enforcement of the Community Reinvestment Act; (b) as a contributor to fact finding, through enforcement of the Home Mortgage Disclosure Act and in the case of the City of Chicago through the enforcement of the city's own more extensive ordinance which also requires disclosure of savings deposits and commercial lending patterns; (c) as an informal broker in negotiations, as was the case with the mayor of Chicago; and (d) as a facilitator of the development process, through grant and loan programs, planning assistance, infrastructure improvement, and other forms of subsidy. The interaction of the three sectors—citizen groups, government, and lenders—creates a "private-public-private" decision-making process (Weiss and Metzger 1986b).

The presence of federal regulatory leverage and the existence of a mobilized political constituency in the form of a coalition of neighborhood organizations were vital to the outcome of the Chicago case. The application of this model of negotiated development to the realm of technological innovation brings with it different actors, goals, strategies, tools, and resources. In Pittsburgh, recent community planning efforts for technological development illustrate the dynamics and possibilities of community collective bargaining as applied to high tech.

High-technology planning in Pittsburgh has centered on the Oakland neighborhood, a racially and ethnically diverse working-class community that is located four miles east of Pittsburgh's "Golden Triangle" of downtown corporate headquarters. The geographic concentration of educational institutions within Oakland, in particular the University of Pittsburgh (Pitt) and Carnegie-Mellon University (CMU), has made the neighborhood a focal point for advanced technology development. Plans and proposals for regional economic revitalization have cited the region's universities as crucial actors in the development process, working in tandem with business and government to develop and transfer technologies to industries and to foster the creation of new technology-based firms (ACCD; Caliguiri et al.). In Oakland, the "critical mass" of CMU's robotics and computer-science expertise prompted the U.S. Department of Defense to locate the Software Engineering Institute (SEI) near the CMU campus. In addition, CMU, Pitt, and other actors have converged around the planned redevelopment of an abandoned Jones and Laughlin steel mill, located within Oakland on the banks of the Monongahela River, into a technological and industrial park (Urban Land Institute).

The political and economic history of Oakland has been characterized by tension arising from the uneasy relationship between the area's institutions and its residential neighborhoods. Oakland's institutions include CMU, Pitt, the six affiliate hospitals of the University Health Center of Pittsburgh (UHCP), and the Carnegie Institute. CMU and Pitt are the two largest universities in the region, while UHCP is a regional health care center and the Carnegie Institute operates major art and natural history museums and related cultural activities.

Situated on a plateau north and east of the Monongahela River, Oakland initially developed as a nineteenth-century suburb. Institutional and residential development in the community proceeded relatively free of conflict in the first half of the twentieth century, and

Oakland enjoyed some of the finest architecture and landscaped urban parks in Pittsburgh. Land-use conflict emerged after World War II, with the proliferation of institutional expansion plans which threatened to alter the shape and dynamic of the community. The conflict initially revolved around Pitt, which initiated a building boom during the 1960s when it changed from a private to a state institution. The number of buildings on campus rose from 23 to 40 after 1964. Pitt also cleared away land in Oakland, as the campus grew from 64 to 110 acres, through the relocation of the Pittsburgh Pirates from Forbes Field and the acquisition of old residential and commercial blocks. By the end of the decade, Pitt embarked on a master plan to construct several large new facilities, involving the purchase of real estate throughout Oakland, often at exorbitant prices.

The population of Oakland began to slip in the 1970s due both to the displacement of residents and small businesses with institutional expansion and to the overall loss of manufacturing jobs in the city. The additional employment generated by the universities and the hospitals has attracted newer, middle-income residents, but only in North Oakland. With some subdivision of low-income units in West Oakland, the community's total population has reached nearly 25,000. The institutions and businesses in Oakland now employ over 20,000 persons, composing nearly 15 percent of Pittsburgh's total employment base.

Many of the community-based organizations within Oakland were formed during the late 1960s and early 1970s in response to the institutional expansion plans of Pitt. The fear of displacement prompted the creation of People's Oakland in 1970. The efforts of this group and other community organizations in Oakland halted the full implementation of Pitt's master plan, and by 1972 a broad public-private vehicle for community planning, called Oakland Directions, Inc., had been established. A comprehensive community planning process from 1976 to 1980 produced the Oakland Plan, which laid out an ambitious agenda for housing rehabilitation, neighborhood improvements and better transportation in Oakland (Oakland Directions, Inc.).

One of the direct spin-offs of the Oakland Plan was the creation in 1980 of the Oakland Planning and Development Corporation (OPDC). OPDC has emerged as one of the most sophisticated community development corporations in Pittsburgh by developing over 120 units of low- and moderate-income and elderly housing, improving the traffic flows and relieving density in Oakland. Governed by a community-based board and funded by the city and private foundations, OPDC has sought to preserve neighborhood stability through community planning, housing and, now, mixed-use development.

The Oakland Plan of the late 1970s did not anticipate the emergence of the Oakland neighborhood as a key growth node in regional economic development. The economic upheaval experienced in the Pittsburgh metropolitan area from 1979 to 1983, when 120,000 jobs were lost—many in heavy manufacturing industries such as steel—prompted the region's business, political, and institutional leadership to stress the role of high technology industries and local universities in Pittsburgh's economic transition and recovery. The geographical concentration of educational institutions within Oakland has made the community a logical target area for technology development activity.

CMU has a strong national reputation in computer science, robotics, magnetics, and special materials. Its research volume has been growing rapidly, rising by 50 percent over the last two years, with the Department of Defense as the largest research sponsor. Forty percent of CMU's research is related to industry, undertaken by either the Robotics Institute or the Mellon Institute. The Robotics Institute, which handles contract research, is growing and has a current budget of $10 million. One-third of its work is for DOD, and it has twenty-five industrial sponsors, including Westinghouse and Digital Equipment. About half of the Institute's work has a clear, mid-range, applied industrial focus, while the other half is general robotics research.

The Department of Defense's Software Engineering Institute (SEI), located in Oakland near the CMU campus, is the first federally funded research and development center to be established in over twenty years. It is expected to create 250 jobs with an annual budget of more than $30 million. The primary audience for SEI will be defense contractors. Sixty percent of its work will be technology transfer to contractors who are expected to open local offices to develop relationships with this research institute.

Pitt's research and development activity is focused on biomedical technology in association with UHCP. Most of this research and development is done on-campus, although UHCP has just developed a Nuclear Magnetic Resonance research institute off-campus in Oakland and is planning a major new Cancer Institute at an undetermined location.

This activity of CMU and Pitt has converged on the planned redevelopment of the abandoned Jones & Laughlin steel mill into an advanced-technology industrial park. The Urban Redevelopment Authority (URA) is the owner of the Oakland site, while the Regional Industrial Development Corporation (RIDC) is the quasi-public developer for the site. CMU hopes to develop a National Center for

Robotics in Manufacturing in the new industrial park. The center would be linked to the work of the Robotics Institute, but would focus more on product development and the production process and would attract spin-off software development firms to the park. Pitt's interest in the park revolves around a biotechnology manufacturing center which would foster the growth of medical research and development activity on the site.

The high technology "boomlet" in Oakland and the redevelopment plans for the J & L site prompted OPDC to assess and define its role in local economic development. Without any role in the decision-making process, OPDC would be able to make no contribution to the high technology agenda created by entrepreneurs and university and government officials for Oakland. Several points of political leverage existed for OPDC that could facilitate access to power and lead to a brokering role for OPDC in the planning and development process.

First, OPDC had an established, working relationship with the administration of Mayor Richard Caligiuri because of its track record as one of Pittsburgh's strongest and most capable community development corporations. In particular, OPDC was an important member of the Mayor's Oakland Task Force. Second, OPDC had developed a working relationship with Pitt as a result of Pitt's support of and involvement in Oakland Directions, Inc., and the implementation of the Oakland Plan. These points of leverage enhanced OPDC's power in two areas—local government funding and regulatory support for high technology development projects in Oakland and the negotiation of a development agreement between Pitt, CMU, URA, and RIDC for the J & L site.

To enhance their bargaining power, OPDC sought to become recognized as an informed actor in high technology development. The absence of a rigorous analysis and projection of the effects and outcomes of technological innovation in Pittsburgh and Oakland provided OPDC with the opportunity to establish its credentials. OPDC obtained a planning grant from the Economic Development Administration of the U.S. Department of Commerce and hired economic development consultants from the University of Illinois at Chicago to prepare an impact assessment of advanced technology development in Oakland (Weiss and Metzger 1986a).

The forty-page impact assessment, completed in early 1986, identifies four key sectors of advanced technology development in Oakland—computer software, robotics, medical research and development, and business support services. Using local and national economic data to analyze trends, the report cites computer software as experiencing fast growth rates in Oakland and Pittsburgh, although the actual number of software jobs as a proportion of the area's total employment base will likely remain small. Software development has mostly produced jobs for keypunchers and computer operators. Oakland's location should continue to make it an attractive site for software firms.

According to the report, the emergence of the robotics industry in Oakland is strongly tied to the continued growth of Pittsburgh's computer software industry. The local market for the application of robotics and software technology will receive a boost from rising Department of Defense expenditures as well as from Pittsburgh manufacturers seeking to modernize aging capital equipment. Industrial retention through robotics development, however, will likely produce job displacement through factory automation that will not be offset by new software jobs.

The report states that medical research and development will continue to exhibit potential for strong growth in Oakland because of local university resources in biotechnology and software. Again, the number of jobs directly generated by medical technology will likely not be large, and the importance of the sector will be more in the commercialization and application process. The indirect effects on employment of advances in research and development are not clear, although the demand for health technicians and software professionals in Oakland will probably grow.

The report projects that the expansion of advanced technology development in Pittsburgh will generate growth in related business support services. High-tech business services will most likely locate within and adjacent to the core of current development in Oakland, creating pressure for office space and increasing local, white-collar professional employment. The scope of technology development and the dynamics of the small office market will set the limits to the growth of business support services in Oakland.

The consulting report assesses the effects on the community of advanced technology development in Oakland in three areas—employment, housing, and office space. Employment will grow, particularly in software, which will be the key sector determining the prospects for technology development in Oakland. The distribution of advanced technology employment in Oakland will be two-tiered. Software professional jobs that are accessible only to those with advanced computer degrees will form the core of the upper tier which will also include a smaller number of highly paid professionals in medical research and development and related business support services. The lower tier will consist mostly of two groups—low-paying, clerical computer keypunching and operating jobs, and technical and medical laboratory support jobs.

According to the report, housing demand by high tech professionals will affect Oakland only if new or renovated housing is developed in sufficient scale to create a new, upscale housing environment. Without programs to ensure affordable new or rehabilitated housing, Oakland is likely to experience continued deterioration of commercial and industrial areas, hampering future institutional and advanced-technology growth. The report also projects that growth in business support services associated with advanced technology development will increase the demand for office space in Oakland. Office demand should be intense near the SEI site, but the limited amount of office development sites should constrain the growth of office space in Oakland's business districts.

The report also examines two crucial high technology development sites in Oakland—the Pittsburgh Technology and Industry Park, and the Software Engineering Institute. The industrial park, located on the old J & L steel mill site, may be delayed by initial development and funding problems. In addition, unless the site's unique urban advantages are accentuated, the widespread availability of good, suburban industrial real estate will hamper leasing efforts, even if the public sector absorbs high initial development costs. The report sees brighter prospects for the Software Engineering Institute. Employment spin-offs from SEI's defense-related research will result from software firms and business support services that locate near the facility. Steady growth in defense spending will help SEI overcome any early difficulties, and the Institute should make an impact in Oakland through job creation for software professionals and increased demand for local office space.

The impact assessment makes five recommendations. First, special services, on-site amenities, public access, and research links with the universities should be promoted on the J & L site through a cooperative process to distinguish it from the traditional model of privately developed, suburban industrial parks. Second, job linkages for Oakland residents should be established within the advanced-technology sector, and employment training provided to equip residents with skills for jobs in high tech. Entrepreneurs, public and private agencies, local institutions, and OPDC should work together to formalize these arrangements.

Third, local institutions and private developers should invest directly in OPDC's efforts to improve the community's housing stock. A housing trust fund financed by institutional and private contributions should be created to support low and moderate income housing development in Oakland. Fourth, OPDC should engage in office development in Oakland so that it may broker for job commitments and generate an income stream to support its nonprofit activities. Finally, the report recommends that the role of Oakland Directions, Inc., and the Mayor's Oakland Task Force in mitigating the impact of development and monitoring growth should continue to be recognized and supported.

The impact assessment has provided OPDC with a tool to establish credibility with the public, private, and institutional actors involved in the high-technology planning process and to inform and organize its constituency of community-based interests around local economic development policy. In the latter arena, OPDC has worked with the North Side Civic Development Council, which also represents a neighborhood affected by advanced technology development, to convene a citywide coalition of community development organizations in Pittsburgh around developing an alternative strategy to the regional economic development plans created by corporate-dominated groups.

The Pittsburgh case outlines a bargaining process over high-technology policy initiated by a community-based group. OPDC has been able to draw upon its own political resources to create a framework for negotiations and to legitimize its role as an important actor in local economic development. Although this case continues to unfold and its final outcome is not yet clear, the information provided by the impact assessment has enhanced OPDC's power and its capacity to organize for broadly based representation in the planning process. This points to the value of institutionalizing a "technology impact statement" into community planning efforts for high-technology development. The impact statement could operate as government-enforced leverage, similar to the federal legislation in the Chicago case, that would contribute to fact-finding and stabilize the bases of power between community interests and private corporations in formulating high technology policy.

The Pittsburgh experience indicates the possibilities of using the model of community collective bargaining to shape high technology development. It also raises several issues that set limits to community planning for technological development. The first group of issues concerns the technological impact statement—it may be difficult to assess the real consequences of high-technology on communities. In particular, the complexity and dilemmas of the employment issue—what kind of jobs are created, who gets the jobs—may be difficult to predict and resolve (Weiss 1983).

The second set of issues focuses upon representation in the community bargaining process. Defining a "community"—its size and its leaders—that is independent of government representation can be difficult amidst conflicting interests and groups. Community representation may lose its relevance and accountability once it is institutionalized into a decision-making structure. Despite its limits, the model of community collective bargaining can be useful to community groups, government officials, workers, and corporate leaders working to initiate a democratic planning process for the introduction of new technologies into their communities.

Works Cited

ALLEGHENY CONFERENCE ON COMMUNITY DEVELOPMENT (ACCD)
1984   *A Strategy for Growth. Vol. 1: An Economic Development Program for the Pittsburgh Region. Vol. 2: The Task Force Reports.* Unpublished document.

CALIFORNIA COMMISSION ON INDUSTRIAL INNOVATION (CCII)
1982   *Winning Technologies: A New Industrial Strategy for California and the Nation.* Unpublished document.

CALIGUIRI, MAYOR RICHARD S., ET AL.
1985   *Strategy 21: Pittsburgh/Allegheny Economic Development Strategy to Begin the 21st Century.* Unpublished Document.

GOODMAN, ROBERT
1979   *The Last Entrepreneurs.* New York: Simon and Schuster.

OAKLAND DIRECTIONS, INC.
1980   *The Oakland Plan: A Citizen's Planning Process, 1977–1979.* Pittsburgh: Urban Design Associates.

PELTZ, MICHAEL, and MARC A. WEISS
1984   "State and Local Government Roles in Industrial Innovation," *Journal of the American Planning Association* 50: 270–79.

URBAN LAND INSTITUTE
1985   *Technology and Industry Park: An Evaluation of Development Potential and Strategies of the Urban Redevelopment Authority of Pittsburgh.* Panel Advisory Service Report. Unpublished Document.

WALSH, ANNMARIE HAUCK
1978   *The Public's Business.* Cambridge: MIT Press.

WEISS, MARC A.
1983   "High Technology Industries and the Future of Employment," *Built Environment* 9: 51–60.
1984   "Dilemmas of State and Local Economic Development Policymaking." Unpublished paper presented to the 26th annual conference of the Association of Collegiate Schools of Planning, New York, NY.

WEISS, MARC A., and JOHN METZGER
1986a   *Impacts of Advanced Technology Development in the Oakland Neighborhood of Pittsburgh.* Unpublished report prepared for

the Oakland Planning and Development Corporation, under a grant from the U.S. Department of Commerce, Economic Development Administration.

1986b "The Role of Community-Based Organizations and Financial Institutions in Negotiated Development." Unpublished paper presented at the conference, Negotiated Development: The Challenge of New Partnerships, Lincoln Institute of Land Policy, Cambridge, Mass.

PART FOUR

Politics

11

Republican Virtue in a Technological Society

ALBERT BORGMANN

Our political climate is marked by the crisis of liberalism.[1] Liberalism
has at least two different levels of political reality, but it is in a critical
state at both of them. At the more superficial level, liberalism is one of
the two contemporary currents in American politics, the one that is
associated with the Democratic party. Let me call it party liberalism.
Its critical malady began with the defeat of Jimmy Carter by Ronald
Reagan in November of 1980.

Liberalism at the deeper level constitutes the dominant view of the
common order in the Western democracies. Springing from many and
varied sources at the beginning of the modern era, it emerged clearly
and powerfully in the nineteenth century and now constitutes the decisive political orientation. Its major achievement was the definition of
the relation between the individual and society. In fact the liberal definition for the first time articulated what we now commonly understand by individual and society. It has become second nature to us to
think of the individual as the first and foremost element of society and
of individual rights as the foundation of the common order. We think
of society as the servant of the individual, and like Mill we are inclined to regard this servant with suspicion and to worry about "the
nature and limits of the power which can be legitimately exercised by

1. I am indebted to Daniel Kemmis for helpful conversations on the major issues of
this essay.

ALBERT BORGMANN

society over the individual" (3). But this social configuration is in contrast to the one that has been characteristic of nearly all of human culture prior to the modern period. Traditionally, the identity and the welfare, the principles and the property of a person have been indistinguishable from those of the group in which he or she lived. The change of the traditional arrangement into the modern one has been signaled most memorably by the title of Tönnies's celebrated study, *Community and Society*. The social as opposed to the communal view of the common order is now so well entrenched that even communists and conservatives had to allow for it and have been unable to establish a self-sustaining alternative to it. Let us call this fundamental view of the common order philosophical liberalism.

I want to show that philosophical liberalism, in spite of its programmatic simplicity and plausibility, is an ambiguous and dubious social design. It owes its force and stability to technology, i.e., to a way of life that is characterized by a division into two correlative spheres: the sphere of the vast productive machinery based on science and engineering and sustained by our labor, and the sphere of affluent and unencumbered consumption to which we devote ourselves in our leisure. These two spheres closely coincide with the public and the private realms, and they specify decisively the liberal division between the social and the individual domains. This articulation of the liberal design I call the technological specification of liberalism.

I will further argue that technologically specified liberalism provides an ethically impoverished common order and therefore needs to be morally redeemed. The critics of liberalism sense this moral impoverishment. But in their efforts to cure this malady they usually fail to recognize the technological specification of liberalism. Hence their efforts tend to be deflected from their goal and end up repairing and reinforcing what they had meant to redeem. Repair takes the place of redemption. To see this in detail, we must attend to the critique of liberalism.

Philosophical liberalism has been criticized since its beginning. Toward the last quarter of this century, the neoconservative attack has been gathering momentum. But one would hesitate to speak of a crisis of philosophical liberalism had it not been for the obvious crisis of party liberalism in 1980. The rejection of the leadership of party liberalism by the voters was taken as a popular repudiation of the philosophical liberal vision. The sense of crisis was deepened by the collapse and exhaustion of liberal theory at the party level. The calls to rally around the traditional liberal ideals became few and far between. Liberal theorists like Lester Thurow, who continued vigorously to champion and elaborate the traditional cause, were found to be less and less inspiring. Liberal politicians no longer relied on the liberal theorists to shape the political agenda; party liberalism was reduced to defense and reaction. The momentum of the national debate shifted to the conservative politicians and theorists.

In what way had party liberalism failed? The conservative charge was that it had stifled the vigor of the American people. Domestically, it had fettered personal initiative and aspiration through appeasement and laxity. The broad charge was spelled out in terms of taxes, unemployment, budget deficits, crime rates, military power, international reputation, and so on. This partisan charge connected well with the deeper philosophical claim that the liberal state had betrayed the highest aim of politics: to help its citizens achieve moral excellence.

The crisis of liberalism was to be cured through republicanism both at the party and at the philosophical level. The reform program of the Republican party had four essential points: (1) to strengthen the family; (2) to assert religion in the public sphere; (3) to restore the free enterprise system; (4) to regain a courageous and powerful stance in international politics. The program has failed obviously in its first two points, and it floundered precisely on the continuing strength of philosophical liberalism. In the liberal view, matters of family and religion belong to the individual; the role of the state in these areas is to protect individual discretion, not to shape or override it. The Republican administration has not been able to modify this common understanding significantly.

The economic programs and foreign policy of the Republican party have failed in more subtle ways. In one way they did succeed. Inflation and unemployment have fallen and, after a deep recession, the economy has been growing again. The expenditures for arms have dramatically increased. But it would be difficult to establish that the Republican administration has been able to effect structural changes in the economy so as to open up opportunities for individual initiative and creativity. The economic upturn has not been the accomplishment of a greater number and variety of entrepreneurs. An advanced industrial economy is a gigantic, intricate, and highly organized machinery that progresses through large-scale planning and through scientific and technological innovation. Such planning is always beyond the reach of individual initiative, and such innovation usually is.

Similarly, a military build-up requires funds, logistics, and technology. If any of these elements requires patriotism, it is the commitment of revenues. But it is hard to believe that senators, voting for increases in the arms budget, were moved by the republican fortitude of their constituents. Foreign policy is a geopolitical calculus and military procurement a scientific and technological task. Neither area provides fertile ground for republican virtue.

In short, the partisan republican programs for the economy and foreign policy appear to be innocuous and inconsequential in the face of the technological character of modern life. A cynic might say that the Republicans' troubles are less profound and that partisan republicanism is but the continuation of philosophical liberalism by other means. But I am inclined to grant the prominent conservatives of the Republican party deeper aspirations if not accomplishments. At any rate, we catch here a first glimpse of what I take to be a crucial issue in contemporary political philosophy, the relation between technology and Republican virtue.

But what we have seen so far is certainly less than compellingly worrisome. The appearance of republican powerlessness may be due to the relatively shallow partisan cast of the programs that we have considered. At the level of philosophical republicanism, we find more incisive critics and thoughtful reformers. If their programs were to be implemented, then the technological organization of society may not be the reef upon which the aspirations of partisan Republicans come to grief as I have hinted. This technological order might instead turn out to be the relatively superficial or accidental circumstance that it is commonly taken to be.

The philosophical critique of the liberal order agrees with the partisan critique that liberalism has put the commonweal at risk. However, the division that emerges at the partisan level between aspiration and policy has its counterpart at the philosophical level as a hidden ambivalence of aspiration. The uneasily stirring question is whether a profound republicanism should aspire to repair or to redeem the common order. Repair is generally called for when something is evidently broken or neglected and will break down altogether if left unattended.

The need for redemption is much more difficult to determine and, above all, to convey. Our society may well appear to be stable and vigorous from the secular point of view, in all the senses of "secular," and we may yet be deeply dissatisfied with it. But to what standards should we resort to justify our dissatisfaction? One hesitates simply to apply the religious or republican standards of excellence, not just because they fail to command widely shared allegiance but, more importantly, because they have an uncertain purchase on most people's daily endeavors. It is not so much that our common activities violate the standards as that they seem to fall altogether outside the scope of those norms.

The difficulty is illustrated by the work of two influential critics of liberal philosophy, Michael J. Sandel and George F. Will. Both have done much to expose the exhaustion of liberalism and to call for a fruitful alternative. Both of them, however, appear to be uncertain that the case for redemption of the common order can be made convincingly; and at crucial points they fall back on the need for repair. In "The Procedural Republic and the Unencumbered Self," Sandel traces the morally dubious course that the relation between the state and the individual has taken in this country. But to the moral indictment he adds another charge: ". . . despite its unprecedented role in the economy and society, the modern state seems itself disempowered, unable effectively to control the domestic economy, to respond to persisting social ills, or to work America's will in the world" (Sandel 1984b, 92).

Will's book *Statecraft As Soulcraft* is dedicated to showing "that liberal democratic societies are ill founded" (18). They have pretended, so the argument goes, that statecraft should not and need not be concerned with the citizens' moral welfare. But statecraft is inevitably soulcraft (19). To ignore this irrecusable task is to perform it poorly. "The result is a radical retrenchment, a lowering of expectations, a constriction of political horizons" (43). Civic virtue declines. "Our sense of citizenship, of social warmth and a shared fate, has become thin gruel" (45). But Will seems to be unsure of the reader's distress in the face of this moral calamity and on the next page adds this warning: "A nation—a civilization—so constituted cannot long endure" (46). In fact, the entire book ends on this note of urgency (164–65).

Of course the need for the repair and the need for redemption of the common order do not have to be disjoined. Outward ruin may be the sign of inward decay. But as in modern medicine we often repair persons without redeeming them, so we have often repaired social ills without curing the moral problems that were at stake in them. In fact the substitution of repair for redemption is the avowed policy of the technological fix, advocated by Alvin Weinberg. A water shortage, Weinberg points out, can be seen as a moral problem that could be largely overcome through neighborliness and thrift, through the virtues of charity and temperance traditionally speaking. And the social engineer (Weinberg's term for the prophet or the teacher of ethics) insists on just this approach. But such moral redemption, Weinberg says, "is difficult, time-consuming, and uncertain in the extreme" (26). Moreover, who gives the social engineer the right to judge and to change people's morals? It is better then to see the water shortage as a technical problem to be fixed or repaired through a technology that provides abundant and cheap water (26–27).

Once more the conflict between technology and republican virtue comes into view here but with a new and significant complication.

When earlier we considered the fate of the partisan republican program, it appeared that one part foundered on the continuing strength of liberalism while the other was repelled by the autonomy of the technological order. Now there is at least the suggestion that republicanism is challenged by an alliance of liberalism and technology, that technology may step into the breach left by the weakening and retreat of liberalism.

But whatever the complexion of the liberal-technological order, has not the ascendancy of partisan republicanism shown that there is a deep-seated popular dissatisfaction with the current order? And whatever the frustrations that partisan republicanism has suffered in its attempt to oblige that dissatisfaction, is it not plausible to think that a more thoughtful philosophical republicanism might engage the doubts and hopes of the people?

For better or worse it appears that the American people's liberal leanings, so obvious since World War II, are as strong as before. The shift of allegiance to the Republican party was apparently motivated by the voters' belief that Reagan would be better able to revive the economy and promote an improved standard of living. The desire for increased economic well-being overrode the common liberal inclinations (Ferguson and Rogers; "Pocketbook Voting").

If this analysis is correct, it exacerbates the difficulty of the republican reformers' task. The voters have given republicans an opportunity to reshape the common order. But the mandate is tied to the condition that overall prosperity increase. Since the beginning of the nineteenth century, the mainspring of growing affluence has been modern technology. And we have seen suggestions that technology and republican virtue are incompatible.

We must act on these suggestions and consider more directly what republican or civic virtue is and what technology is. The first thing to notice is the prominence of the call for republican virtue. It is one of the two central concerns of the widely discussed *Habits of the Heart* (Bellah et al.). In one way or another it is ever on the lips of George F. Will. It is the concluding plea of another influential book, *After Virtue* (MacIntyre). And a recent survey of work on virtues calls for more analysis of it (Pence).

What is just as striking are the variety and vagueness of these calls. What is here called republican virtue is elsewhere called public or civic virtue, civility, loyalty, communitarianism, or the republican tradition (Bellah et al. 254–56, 270–71; MacIntyre 244–45; Oldenquist; Sandel 1982, 179; Sandel 1984a, 17; Sandel 1984b, 90; Spreng and Weinberg 38–39; Will 134–35). Of course these terms do not all mean exactly the same thing. But the vagueness with which these terms are used both suggests a common concern and makes it impossible to set them off from one another in a systematic way.[2] Compared with the elaborate systems and discussions of justice, the treatment of republican virtue is scattered and inchoate. There appears to be a common intuition, but there is little sense of a principled and shared scholarly endeavor. A look at the historical background can help us to understand why the issue at hand is so elusive or recalcitrant.

There are established categories of virtues in philosophy. We understand what is meant by the moral, intellectual, cardinal, or theological virtues. But one will look in vain for a similarly clear and firm category of republican or civic virtues. Ironically, the cardinal virtues—temperance, fortitude, wisdom, and justice—come out of Plato's *Republic* and were conceived as social or political virtues. Similarly, Aristotle's moral virtues were the habits of excellence that distinguished the Greek citizen in his commerce with his fellows and in his concern for the city.

Initially, then, the cardinal and moral virtues were of themselves civic virtues. But they existed at a critical distance to the social order from the start. They are the standards of excellence that we associate with Athens. But at the time that they were formulated, the unity, the power, and the splendor of the civic culture at Athens were already fractured. In Rome, the Greek civic virtues were never fully at home. In fact, "republican virtue" is an eminently Roman term. It commemorates the stern and self-sacrificing citizens who founded the Roman republic in the period between the original kings and the later emperors. Virtue, in their case, was not a generic term for several species of excellence but a singular and comprehensive attitude of severity and service.

During the high Middle Ages, Thomas Aquinas, in the second part of the second book of the *Summa Theologica*, subordinated the moral and cardinal virtues to the theological ones (faith, hope, and charity) and integrated the former with numerous other virtues and vices. Thus republican virtue and the Greek civic virtues lost their political distinction. Moreover, both the classical and Christian virtues were in some conflict with the heroic virtues of the Germanic feudal tradition that lived on in the notion of chivalry and terminated late in the Middle Ages in the ideal of loyalty.

Briefly, at the beginning of the modern period, civic virtue attained conceptual distinction in the writings of Montesquieu, who set political virtue apart from moral and Christian virtues (538, 822–23). Republican virtue became practically significant in the founding of this

---

2. An exception to the general vagueness is the treatment of loyalty by Oldenquist.

country, though it is a matter of great controversy just how influential it was. At any rate, Jefferson's initial devotion to and gradual despair of a republic of independent farmers and artisans is a commonplace of American history.

I believe that the troubled and varied fate of republican virtue constitutes a positive phenomenon. Republican virtue escaped the professional purification of standard ethics and remained a topic of the wider and more consequential conversation of humanity (Rorty 1976; 1979). Even now it is most thoroughly discussed by cultural and political historians, not by professional philosophers (Kasson; Pocock). The history of republican virtue is therefore richer and more suggestive than that of ethics. At the same time, the teachings of republican virtue have escaped the fate of ideology; they have never been pressed into the service of justifying or excusing the failures of an existing social order (Ricoeur).

Talk of republican virtue, therefore, can have a uniquely fruitful and critical force today. It can help us to recover a proper obedience to the past and to overcome our crippling obsequiousness to the present. Obedience to the past is not a copying of past models, but the ability to listen to the voices of history and to allow oneself to be challenged by the aspirations of the republican tradition. From such a conversation with history we may draw the strength to question the obsequious common allegiance to the prejudices of the contemporary social order.

Let me now return to the contemporary discussion of republican virtue to show how it contains the seeds of a fruitful and critical approach to the common order. In spite of the variety and vagueness of the discussion that I noted earlier, a common aspiration is detectable. Most abstractly put, it is a desire for selflessness and social cohesion. Selflessness is rendered as loyalty, benevolence, fraternity, social sympathy, generosity, commitment, and in similar terms. Social cohesion is spelled out as common purpose, public ends, public good, community, and in other ways (Bellah et al. 168–77, 187–90, 251–52, 335; Oldenquist 177; Sandel 1982, 32–33, 169–70, 179; Sandel 1984b 90, 93, 95; Will 134–35). These two ideals are thought to be intrinsically good but imperiled or lacking in the liberal state.

The philosophical republican program does not remain at this level of abstraction. But neither is it spelled out with a concreteness sufficient to resist misunderstanding and subversion. To simplify matters let me point out the problem at the abstract level. It is natural to construe the republican philosophy the following way: Without selflessness, no cohesion; without cohesion, no survival or, at any rate, no prosperity. But the advanced industrial societies, ours in particular, have shown at least since World War II an astounding resilience and productivity. The inferential chain contained in the common construal of philosophical republicanism is but another reflection of the fallacious attempt at resting the need for redemption on the need for repair. To restate the issue in the philosophical terms of the republican aspiration: If selflessness is merely instrumental to social cohesion and cohesion to survival or prosperity, selflessness can be dispensed with, and cohesion can be given a nonrepublican interpretation. It is the pursuit of prosperity that constitutes the common purpose. That purpose is much more efficiently served by scientific technology than by selflessness. The only loyalty or commitment that has to be required of the citizens of the technological society is to the system of production and consumption. Such loyalty is simply the willingness to support the technological structure through labor, taxes, and through compliance with the rules of the game. That sort of loyalty has never been perfect but, still, it has been sufficient to maintain and expand the technological order (Borgmann 101–13).

We have seen before that the republicans, both the partisan and the philosophical ones, set out to attack liberalism but end up being defeated by technology. The reason is their failure to recognize the technological specification of liberalism. Mill, to take a prominent example, sought to restrict the moral authority of society in order to make room for the autonomy of the individual. We commonly fail to realize, as did Mill himself, how vague and even incoherent this program is. The program, as it stands in *On Liberty*, could have been given a libertarian, an anarchical, or neolithic realization (Mill). What makes Mill's proposal so plausible is its consonance with the technological order. The autonomy of the individual is acted out in the freedom of choice of labor and consumption. Society is restricted to the construction and maintenance of the productive machinery.

The philosophical republicans recognize that a moral life that is divided between solitary choices at the individual level and impersonal productive engagement in a society of hundreds of millions is fatally impoverished. But they lay the blame for this calamity at the door of liberalism. In this way they not only overlook the concreteness and power of the present social organization but also the source of its vitality. Political rhetoric can easily turn people against partisan liberalism; but it has yet to budge people from their allegiance to technology. Here we begin to understand why partisan republicans have been able to displace partisan liberalism only to preside over the continued expansion of philosophical liberalism, enacted technologically. In doing so they have repaired some of the flaws of the liberal technological rule; but they have not even begun to redeem it.

ALBERT BORGMANN

The partisan republicans' ignorance of technology has blinded them to the magnitude of the task of redemption. Let us consider the problem of decentralization to illustrate the enormity of the challenge. The cohesion of something so large and complex as a modern society requires—and is, in fact—centralization. Hence to undo the morally debilitating simplicity of the society-individual schema requires, so it seems, that we break up this centralized structure. Decentralization is therefore properly dear to the republican reformers (Bellah et al. 45–46, 152–63, 292; MacIntyre 33, 68, 181–83, 244–45; Oldenquist 192–93; Sandel 1984a, 17; Sandel 1984b, 91–95).[3] To advance the cause, they often employ the rhetoric that denounces bureaucratic and corporate arrogance. But such criticism remains at the surface of the problem. Centralization is not ultimately the arrogation and configuration of political and economic power; it is rather a requirement of leisure and consumption. Daniel Spreng and Alvin Weinberg have pointed out that if one wants to maximize or save free time, one must expend a maximum of energy and expertise (137–39). An individual of necessity commands limited expertise and energy. To take Spreng's and Weinberg's example, a radically decentralized energy system requires both technical expertise and attention of the individual owners. A highly centralized system can mobilize an abundance of expertise and energy and gives the individuals "the freedom to allocate time" as they wish (139, 141, 144).

Thus, surprisingly to the republican cause, centralization can be seen not as a threat to freedom but as a condition of it. Moreover, it is also the basis of a kind of decentralization. Centralization did not occur, as Langdon Winner has observed, in the area "of the everyday consumption and use of things. In that domain individuals remained the centers of activity. Noting this fact, a number of observers have argued that the automobile is an extremely decentralized form of transportation" (93).

What we see here is the technological specification of the liberal state. The backbone of the liberal society is the centralized system of production and administration. The autonomy of the individual consists in the leisure and affluence of consumption. Let me restate here the sense of technology that I have employed all along. I use technology as the name for the decisive pattern of contemporary culture. That pattern is characterized by the division of common life into the sector of the scientifically grounded productive and administrative

3. Ironically, decentralization is also a social goal of a certain faction of the Left (Winner 89).

168

machinery and into the sector of unencumbered, individualized consumption.[4]

An analysis of technology in the proper sense of the term discloses the center of gravity of the common order more sharply than the conventional political and social investigations. It shows how people are connected to the present state of affairs. In highlighting that crucial point I can do no better than quote Winner at length:

> For most of the twentieth century the prospect of unhindered personal consumption and use of goods seemed to make most people happy; in fact, that is what "freedom" began to mean in the eyes of many. It did not matter that General Motors had become so large and powerful, since that meant that the company produced an inexpensive, reliable car for our weekend spin in the country. It did not matter how electricity was made or carried over a vast electric grid as long as you could flick a switch and have the lights go on. It did not matter exactly what Kellogg was putting in the cornflakes as long as the cereal tasted good and seemed to contribute to strong bones and healthy bodies. A primary source of legitimacy for many of the systems that form the heart of the technological society was that consumption was still centered in the individual. Any notion that ordinary people might want to have control over production or have a say in decisions beyond those of the immediate enjoyment of goods and services seemed out of the question. (93)

The technological orientation specifies, if it does not supersede, the official political organization of the Western democracies. Spreng and Weinberg show well how politics is assimilated to technology as a way of life.

> It is not self-evident that people, by and large, wish to be bothered by what they perceive as distractions from their own pursuit of satisfaction, or for that matter, that the uniquely noble way to spend our lives is through intimate involvement with government, not to say with our personal Stirling engine or solar collectors. Indeed, as Robert A. Dahl says, "To be interested in politics . . . need not compete with one's primary activities. By contrast, active political participation frequently removes one from the arena of primary activities. Since the primary activities are voracious in their demands for time, political activity must enter into competition with them. For most people it is evidently a weak competitor." Apparently most people would, if given the choice, rather perfect their backhands or their piano-playing

4. This pattern has many more aspects and implications (Borgmann).

ALBERT BORGMANN

than actively confront the intricacies of public policy or recalcitrant prime movers. (142)

Winner similarly says: "One delegates power and authority to representatives, to bureaucrats, to the President, or to other such distant
persons to get the business of government out of your way. These are
not matters for which ordinary citizens want to be responsible" (95).

This is what republican virtue is finally up against. If it is to prevail,
its proponents must accomplish two tasks. The first is the translation
of the critique of liberalism into a critique of technology. Once the task
is seen, it does not pose any insuperable obstacles. In fact scattered
pieces of that more incisive critique are already on hand in the writings of the republican philosophers (Bellah et al. 41–44, 50, 72–73,
256, 261–62, 281, 284; MacIntyre 33, 68; Oldenquist 189–90, 192–93;
Sandel 1984b, 92–95).

The other and more daunting task is to give an account of republican selflessness and cohesion that is concrete and, most importantly,
stands in a fruitful and affirmative relation to technology. It is the decisive task because without at least some credible beginnings of a
more engaging and ennobling vision of the common order, people are
unlikely to let go of their allegiance to the present order. Happily, here
too one need not begin from nowhere. Again it is a matter of gathering and orienting various developments and proposals. Let me con-
clude with a sketch of republican virtue in a technological setting. The
task is to give the republican intuitions of selflessness and social cohesion firm and vital embodiment.

The search for embodied selflessness must take its clue from the
character of technological selfishness. The latter consists in the seemingly splendid isolation of one who is entirely insulated from the forces
of nature and from the claims of culture. Selflessness is achieved not
when one sacrifices oneself to abstract principles, but when one gen-
erously responds to natural circumstances and cultural traditions and
so overcomes self-centeredness (Oldenquist; Toulmin). Genuine community can only grow out of that kind of generosity.

To open technology to traditional forces is not to break up all the
central structures of technology. Nor is it to conserve yesterday's artifacts and rituals and to transport them in an embalmed state, as it
were, into the present. It is rather to secure within technology a central space for things and practices that are continuous with the past
but are now allowed to flourish in their own right.

In this way we must open technology for good work and celebration. On the side of work this is to promote and secure through political measures a dual economy, a local labor-intensive industry of arts

170

and crafts alongside the centralized and increasingly automated sector. To affirm the dual economy politically is to work out and to fix through legislation a reasonable division of labor and production between these two sectors, to promote the centralized economy through the cooperation of business and government, and to secure the welfare and vitality of the local economy through tax, credit, and social service statutes that involve a minimum of bureaucracy and regulation. A decentralized economy that engages our physical and communal gifts and responds to its natural setting and cultural circumstances provides fertile ground for the republican ideals of which Sandel speaks so eloquently, for "those loyalties and convictions whose moral force consists partly in the fact that living by them is inseparable from understanding ourselves as the particular persons we are," for "enduring commitments and attachments," and for "the expansive self-understandings that could shape a common life" (Sandel 1984b, 90–91).

A reform of work and production along these lines raises a number of difficult questions. I have tried to answer them elsewhere (Borgmann 237–41). Let me here turn to the question of how we can make the technological setting generous to communal celebration. Generally speaking, it is a matter of providing public places for communities of celebration. By celebration I mean the regular and active devotion to a tangible cause that is its own reward by a fair number of people who come to know and appreciate one another through their common engagement. Celebration in this sense begets community. To provide a public place for celebration is literally to secure a prominent space for it in or near a city along with the personnel needed to maintain it and to do so through public revenue and under the supervision of the government.

More particularly, by celebrations I mean athletic, artistic, and religious practices. Clearly the republican endeavor to support such practices publicly clashes with both partisan and philosophical liberalism. John Rawls, for example, allows for the support of such concerns through the so-called exchange branch of the government, but only on condition that there be near unanimous popular consent to such measures (282–84, 331–32). So strong a requirement is as good as a constitutional prohibition. I think the liberal attitude is inappropriate on this point. But it still deserves a careful reply.

To begin with current practice, we already provide partial or full public support for athletic practices. There are municipal tennis courts, swimming pools, golf courses, baseball fields, and other facilities. Surely, the necessary expenditures fail to have unanimous support. There are academicians, musicians, and balletomanes who would dissent. We do support some of the arts as in the case of public museums or municipal concert halls.

From the strict philosophically liberal point of view, the support of athletic or musical excellence is no more defensible than that of religious devotion. Nor can we fall back on an argument from popular demand. The U.S. Bureau of the Census states that there were about 14.3 million golfers in the United States in 1983; but there were roughly 52.1 million Roman Catholics (51, 227).[5] Moreover the strict enforcement of the policy of philosophical liberalism would be impossible. We might close down all public facilities that people use to celebrate life, including parks and playing fields. And in public education, we might eliminate the sports, the arts, and everything that has more than instrumental value and that in embodying an element of the good life biases students toward that life.

But the resulting social structure would itself institute one enormous bias, viz., one toward unencumbered and individual consumption and against any kind of traditional and communal life. It would also make society still less egalitarian than it is now since only the rich through ownership or exclusive associations would have access to a great number of the settings necessary for the pursuit of excellence.

But more positive factors can be instituted and pointed out to oblige the liberal apprehensions. First of all, no community of celebration should be publicly supported unless it has open membership. A community of celebration cannot exclude anyone on irrelevant conditions such as race. But it has the right to make devotion to its cause a condition of admission. This is a practice we are following now. Anyone can play on a municipal golf course. But you cannot use one of the fairways for your soccer team. Similarly we should build cathedrals for Episcopalians as long as any sincere believer can join the parish. But an orchestra that wants to use the church just to make music cannot do so without the consent of the parishioners.

What now guards our public tennis courts, swimming pools, and parks as places of celebration is a spirit of respect and a principle of self-selection. The soccer players do not invade the tennis courts, and the tennis players stay away from the parks. Sectarianism is also kept within the bounds of civility through the overlapping memberships in the communities of celebration (Oldenquist 177, 179). I may not play golf; but I wish the golfers well because some of them I know as fellow skiers and others as brothers and sisters within my religious commu-

5. I am assuming that the First Amendment prohibits Congress from establishing a state religion; it does not prohibit Congress or state or local governments from supporting various and indefinitely many religions. Needless to say, this is a controversial assumption.

nity. The same spirit of tolerance and interconnected loyalties can sustain a republic of celebrations.

But why not entrust celebration entirely to voluntary associations as liberals suggest (Rawls 328–29)? As I have already mentioned, such a system is inherently undemocratic in being inegalitarian and unrepublican in surrendering excellence to the whims of plutocracy. More profoundly, one might say that celebration needs public policy and public policy needs celebration.

Whoever has struggled to establish or sustain a truly voluntary association, dedicated to secure a place for excellence and celebration, will know the inhuman burdens of such a task. Like Oscar Wilde's socialism, such an endeavor takes too many evenings. And if the enterprise is of any significance, it will be expensive and require fund raising. But in any society, money and power are always spoken for. If the voluntary association is not part of the establishment in disguise, it must assert itself ever and again against inertia and indifference. This leads to rapid exhaustion on the part of board members and officers and to an uncertain life for the association, if not to its demise.

All this shows that genuinely voluntary associations, that is, those that decisively depend on avocational volunteer work, are unsuited for significant social and moral tasks. Indeed, to assign important common matters to voluntary associations is philosophically incoherent.[6] It is to give such matters a place where society has made no room for them. If there is a significant social and moral issue, society must properly, that is, publicly and officially deal with it and give it political affirmation. That comes to giving the issue a firm place within the customary framework of the public budget. Of course public support for communities of celebration is not a matter of all or nothing. The philosophical concern is crucial: to give communal celebration a prominent and secure place in public life. If that concern is publicly agreed on, ways can be explored to find appropriate combinations of private and public support.

Communal celebration, at any rate, deserves and needs the affirmation of public policy. Conversely, public policy needs celebration. As it is, the substance of the good life is excluded from the sphere of government. According to the ruling liberal conception, government is properly concerned with curbs and gutters, but not with excellence and celebration. This misunderstanding gives us the worst of two worlds. It gives our shared concerns a typically dry and technical cast

---

6. Voluntary associations can be taken in a sense that goes beyond avocational and volunteer work. But in that wider or deeper sense they fail to be consistent with the liberal vision for different reasons (Borgmann 97).

that engenders civic apathy. At the same time it fails to gives us a genuinely restrained government, one that is respectful of a profound pluralism; our government is rather the ultimate facilitator and guarantor of the technological way life.

To draw communal celebrations into public policy is to enable the polity to speak and act on what finally matters to us. Undoubtedly there will be risks and controversies in such revitalized politics and a need to proceed cautiously and tolerantly. But there is also the chance of a more communal and cohesive spirit just because we would share with one another our deepest concerns. Finally, philosophically grounded republican politics will answer to what is valid and realistic in the call for decentralization. No bureaucracy in Washington, D.C., can possibly know what kind of celebrations are appropriate in Missoula, Montana, given the land, the seasons, the local traditions and the aspirations in western Montana.

What republican virtue needs is not a concern with moral principles, but with *res publicae*, with the tangible things and practices that will rightfully claim devotion and inspire excellence once we have given them a publicly agreed upon and prominent place in our lives.

Works Cited

BELLAH, ROBERT N., ET AL.
1985   *Habits of the Heart*. Berkeley: University of California Press.
BORGMANN, ALBERT
1984   *Technology and the Character of Contemporary Life*. Chicago: University of Chicago Press.
FERGUSON, THOMAS, and JOEL ROGERS
1986   "The Myth of America's Turn to the Right," *The Atlantic Monthly*, 252: 43–53.
KASSON, JOHN F.
1977   *Civilizing the Machine*. Harmondsworth, England: Penguin.
MACINTYRE, ALASDAIR
1981   *After Virtue*. Notre Dame: University of Notre Dame Press.
MILL, JOHN STUART
1956   *On Liberty*. Edited by Currin V. Shields. Indianapolis: Bobbs.
MONTESQUIEU
1964   *Oeuvres Completes*. Paris: Seuil.
OLDENQUIST, ANDREW
1982   "Loyalties," *Journal of Philosophy* 79: 173–93.
PENCE, GREGORY E.
1984   "Recent Work on Virtues," *American Philosophical Quarterly* 21: 281–97.
_____
1986   "Pocketbook Voting," *Scientific American* 255: 62

Pocock, J.G.A.
1975   *The Machiavellian Moment*. Princeton: Princeton University Press.

Rawls, John
1971   *A Theory of Justice*. Cambridge: Harvard University Press.

Ricoeur, Paul
1979   "Ideology and Utopia as Cultural Imagination." In *Being Human in a Technological Age*: 107–25. Edited by Donald M. Borchert and David Stewart. Athens: Ohio University Press.

Rorty, Richard
1976   "Keeping Philosophy Pure," *Yale Review* 65: 336–56.
1979   *Philosophy and the Mirror of Nature*. Princeton: Princeton University Press.

Sandel, Michael J.
1982   *Liberalism and the Limits of Justice*. Cambridge: Cambridge University Press.
1984a   "Morality and the Liberal Ideal," *New Republic* 190: 15–17.
1984b   "The Procedural Republic and the Unencumbered Self," *Political Theory* 12: 81–96.
1985   "The State and the Soul," *New Republic* 192: 37–41.

Spreng, Daniel T., and Alvin M. Weinberg
1980   "Time and Decentralization," *Daedalus* 109: 137–43.

Tönnies, Ferdinand
1979   *Gemeinschaft und Gesellschaft*. Reprint of 1887 edition. Darmstadt: Wissenschaftliche Buchgesellschaft.

Toulmin, Stephen
1981   "The Tyranny of Principles," *Hastings Center Report* 11: 31–39.

U.S. Bureau of Census
1984   *Statistical Abstract of the United States 1985*. 105th edition. Washington, D.C.

Weinberg, Alvin M.
1986   "Can Technology Replace Social Engineering?" In *Technology and the Future*. Fourth edition: 21–30. Edited by Albert H. Teich. New York: St. Martin's Press.

Will, George F.
1983   *Statecraft as Soulcraft*. New York: Simon.

Winner, Langdon
1986   *The Whale and the Reactor*. Chicago: University of Chicago Press.

12

Brandeis v.
Nineteen Eighty-Four
Computer Technology and
the Right of Privacy

SENATOR CHARLES MCC. MATHIAS, JR.

The evolution of computer technology has transformed the way we live, work, and play. The phenomenal growth in the power and pervasiveness of this archetypal high-tech creation also challenges some of our most basic ethical precepts, concepts of rights and responsibilities that are woven into the fabric of our organic law: the Constitution. One of those bedrock precepts is the right to privacy. I will discuss two aspects of this new challenge to the right of privacy: the threat of new methods of computer surveillance to the privacy of individuals and the need to protect privileged information transmitted through new telecommunications media.

The Individual's Right of Privacy

The right to privacy is not explicitly mentioned anywhere in the Constitution, but it has been inferred by judicial decision from a number of provisions of the Bill of Rights. It was in a case involving the Fourth Amendment—the right to be free from unreasonable searches and seizures—that Justice Louis Brandeis produced a historic definition of the right to privacy. The case was *Olmstead v. United States*, which

An earlier version of this chapter was the keynote address to the conference "The Human Side of High Technology," sponsored by DePaul University's Institute for Business Ethics and Illinois Bell, November 1984.

concerned wiretapping. The famous Brandeis dissent in *Olmstead* is more than half a century old, but it speaks eloquently to our modern predicament. "The makers of our Constitution," Brandeis wrote, "sought to protect Americans in their beliefs, their thoughts, their emotions, and their sensations. They conferred, as against the government, the right to be let alone—the most comprehensive of rights and the right most valued by civilized men."

Brandeis understood that the Constitution protected this right to privacy against all sorts of threats posed by all sorts of technology. It prevented a modern police force from an unwarranted telephone tap as surely as it prevented a redcoat from rifling a colonial's desk. And, he emphasized, it would have to meet similar challenges in the future. As he put it, "The progress of science in furnishing the government with means of espionage is not likely to stop with wiretapping. Ways may someday be developed by which the government, without removing papers from secret drawers, can reproduce them in court, and by which it will be enabled to expose to a jury the most intimate occurrences of the home. . . . Can it be," Brandeis asked, "that the Constitution affords no protection against such invasions of personal security?"

To Brandeis's colleagues on the Supreme Court of 1928, that question must have seemed pretty fanciful. But we know better. If Brandeis's "someday" has not arrived yet, surely it is just around the corner. Before long—thanks, in large part, to the accelerating pace of progress in computer technology—we must confront Brandeis's question.

Today, the computer has become indispensable in commerce, industry, and government. Increasingly, information is conveyed by one computer to another, covering vast distances in seconds. The bank, credit, medical, and business records of almost every American are stored in some electronic memory. Computers do not discard information, unless ordered to. They do not forget it. They amass it, retain it, and produce it indiscriminately, at the touch of a button.

The threat the computer poses to the right of Americans to privacy—the "right to be let alone"—is, in one sense, a familiar one to any student of the humanities. In Shakespeare's *Henry V*, the Duke of Bedford voices the fear that, "The King has note of all they intend / By interception which they dream not of." Today, we understand those interceptions all too well. Bedford's fear has entered our collective unconscious as a waking nightmare. It is encapsulated in the more contemporary literary formulation: "Big Brother is watching you."

Just a short time ago, the approach of the year 1984 pushed George Orwell's modern classic once again to the top of the best-seller lists.

For a brief period the popular press was filled with clashing opinions about how closely or faintly modern America resembled Winston Smith's oppressive Oceania. Many of those essays proved to have a very short shelf life. But one remark stays with me. It came from one of our nation's leading privacy experts, Professor David Linowes of the University of Illinois, who served as chairman of the Privacy Protection Study Commission during the mid-1970s. Professor Linowes observed that "the skeleton for George Orwell's *1984* is already here. All it needs is fleshing out by a Big Brother."

High technology provides the framework for the realization of Orwell's dark vision. The phenomenal increase in computing power, the development of ever more sophisticated surveillance techniques and devices, the linking of data bases into ever more complex networks—from this scaffolding could arise the structures of oppression that overwhelmed Winston Smith's right to be let alone. Whether we will in fact witness what one journalist has called "the rise of the computer state" depends upon our vigilance against attempts to put flesh on the skeleton of *1984*.

I have no doubt that our society would never tolerate the Big Brother of Orwell's vision—a regime of silent, continuous, pervasive surveillance that rips away the veil of privacy. But harder questions cluster around the margins of the problem. The tougher ethical issues arise when the values that underlie our respect for privacy clash with other, equally compelling goals. Let me suggest a few examples.

First, what about computerized surveillance as punishment? Our society is increasingly troubled by violent crime. We are particularly frustrated when criminals emerge from prison to commit more crimes, and when crimes are committed by persons on probation after conviction, or on parole after imprisonment. Computer surveillance could help to break this cycle. A person on probation or parole is already required to adhere to restrictions on his or her movements, associations, and activities. The computer could make it easier to enforce these restrictions.

A pilot project underway in New Mexico demonstrates the system in a primitive form. There, some misdemeanor offenders are outfitted with electronic anklets that transmit an alarm to a central computer if the probationer strays more that 1,000 feet from the telephone. Every morning, the computer prints out a list of the subject's coming and goings on the previous day. It would be a relatively simple matter to add the capability to monitor all the probationer's conversations. Before long, we may see more and more offenders sentenced to continuous surveillance by the state.

A more effective deterrent to the repeat offender could hardly be imagined. The careers of professional criminals would be cut short, at a fraction of the cost of incarceration. And only the criminals would feel the cold eye of Big Brother.

How would our society respond to all of this? After all, a person who has been convicted of a crime can legitimately be deprived of many constitutional rights. Is the "right to be let alone" one of them? Or must this right, which Justice Brandeis called "the most comprehensive and most valued," be preserved, even though others have been forfeited?

Second, what about the growing use of computers to detect fraud in government benefit programs? Everyone agrees that we must remove from the rolls those who are not entitled to food stamps, aid to families with dependent children, or other benefits. A computer match of two files that ought to be mutually exclusive seems like an efficient way to detect fraud. For example, few if any legitimate beneficiaries of AFDC will also be recipients of large amounts of interest income from bank accounts. So why not match the two files and, within minutes, spot the cheater?

It is no wonder that more and more government agencies, at all levels, are turning to computer matching as a key enforcement tool. But like any tool, it is not well suited for all jobs. The quality of the computer match is, at best, only as good as the quality of the data in the two files that are being compared; and anyone who has ever had a dispute with a government computer, whether over a traffic ticket or a tax return, can supply his own horror story about deficiencies in the quality of the data collected by government agencies. In computer matching, as in any other information processing, the rule of "garbage in, garbage out" applies with full force.

Many government benefit programs are targeted at the poorest of the poor—the kind of people for whom a wrongful termination of benefits may mean homelessness, malnutrition, and despair. That is why fairness requires that the futuristic technology of computer matching be coupled with an old-fashioned concept that dates from the days of the Bill of Rights, two centuries ago—due process of law. Those whose names are flagged by the computer must have the right to be heard, and to present their case to a flesh-and-blood decision maker, before drastic action is taken.

Computer matching also challenges some widely accepted principles of information privacy. When Congress adopted the Privacy Act just ten years ago, it incorporated certain standards of "information ethics." One of the pillars of this code of fair information practices is the idea that information collected for one purpose generally should not be used by the government for other, unrelated purposes. This principle is particularly important when the information involved is personal, the sort of data over which individuals ought to be able to assert some degree of control in the name of privacy.

The practice of computer matching is, of course, sharply at odds with this principle. When the Internal Revenue Service collects information about interest income, it does so for the purpose of collecting tax revenues. It is not concerned with policing eligibility for AFDC. To use the data for that purpose conflicts with the principles underlying the Privacy Act. It also erodes one important barrier to the creation of a centralized national data bank on every individual's contacts with the government—a chilling scenario that is right out of Orwell.

Third, the latest proposal to use computers to police the benefit rolls goes beyond computer matching. The Department of Agriculture has announced a pilot project to convert food stamp benefits from a paper coupon system to an electronic funds transfer model. At the supermarket checkout counter, the food stamp recipient's purchases will be tallied by a computer terminal, and the amount of the purchase will be deducted automatically from his "account." This project could go a long way toward curbing several kinds of food stamp fraud. But it could also bring us just a short step away from the creation of a data base that will reveal when and where each recipient used benefits to make a purchase, and exactly what items were purchased. If we let Big Brother oversee the shopping lists of food stamp recipients, we will have strengthened the integrity of this massive benefit program, but at a high cost to the right of poorer Americans to be let alone.

The Protection of Privileged Communications

Most of the examples which we have looked at so far involve individuals who are on the receiving end of computer technology. But that is only part of the story. Thanks to the microcomputer revolution, more and more of us are becoming computer users. The proliferation of the new information processing technology in our homes, schools, and offices raises a host of questions related to the protection of privileged communications.

As more and more Americans become computer-literate, we increasingly use various new communications media. The computer-to-computer data connection supplements—or sometimes replaces—oral communications by telephone. Cellular telephones, cordless telephones, local area networks, electronic mail—these and other new communications media are springing up everywhere. And individuals and businesses are making wide use of these new ways to share information.

Some of the messages that these new media carry are highly sensitive. A translation of the digital blips racing by wire, microwave, fiber optics, and other paths could reveal proprietary corporate data, or personal medical or financial information. The users of these new networks—and that means all of us—expect legal protection against unwarranted interceptions of this communications stream, whether by overzealous law enforcement officers or private snoops.

But the law as it now stands may not provide that protection. Under the 1968 wiretap law, the privacy of our communications may turn on technical questions—whether the communication is carried by wire, whether it is in analog or digital form. These questions are simply irrelevant to the legitimate expectations of those who transmit and receive information in today's communication networks. If privacy law lags behind technology, then our task is to bring the law up to date. The track record in this regard is not particularly heartening. Forty years after the *Olmstead* case, the Congress finally passed a federal statute on wiretapping. If we wait that long to safeguard the new communications links, our privacy rights will have been dealt a serious blow.

Recently we missed one opportunity to plug this loophole. In October 1984 Congress passed the first federal computer crime law. It outlaws, under certain circumstances, unauthorized intrusion into computer data bases; but it ignores the question of interception of communications between those computers. To use a low technology analogy, it is now a crime to steal a letter from my mail box, but not to grab it from the letter carrier before it is delivered. I hope that this anomaly will be corrected through federal legislation.

In this situation, too, there is a countervailing value we must consider. Just as the computer has become an integral part of the operations of a wide range of American industries and commercial institutions, so it has also been exploited in the service of criminal enterprises, large and small. Law enforcement needs new tools to attack this new breed of criminal. The power to intercept computer communications may be one of those tools. But unless the power is confined by a warrant requirement, and other limitations similar to those now applicable to conventional telephone taps, our "right to be let alone" will be needlessly jeopardized.

Conclusion: The Value of Citizen Vigilance

Obviously, the conflict between countervailing values in the selection among the possibilities afforded by the implementation of computer technology raises a series of hard choices. Is it fair to impose upon wrongdoers a pervasive system of surveillance we would never accept for ourselves? Can we root out fraud in government benefit programs without intruding on the privacy of legitimate recipients? Can we fight computer crime without unleashing Big Brother? These are just a few of the ethical dilemmas we will be facing in the new future.

At this stage in the implementation of computer technology, it is too soon to tell whether the skeleton of Orwell's nightmare is being fleshed out. Some interesting polling data reflects the public's ambivalence on this question. A Louis Harris poll, taken for the Southern New England Telephone Company in 1983, reveals that nearly nine-tenths of the public thinks that the development of computer technology will make the quality of their lives better. But at the same time, a slim majority of the respondents believes that the *present* uses of computers are an actual threat to personal privacy in this country. Interestingly, the percentage expressing this viewpoint has increased steadily over the years; in 1974 barely one-third perceived a threat to personal privacy.

I suppose that we can look at this glass as either half full or half empty. On the one hand, perhaps this trend simply mirrors the technological reality: Justice Brandeis's vision of insidious surveillance is now a genuine possibility, not a pipe dream. But on the other hand, it may be a healthy sign that the American public is becoming more sensitive to the threat to its right to privacy. It is also reassuring that this wariness is not based on ignorance. After all, the American public today is far more conversant with this technology than it was a decade ago. This familiarity seems to have brought a heightened appreciation of the potential dangers, as well as benefits, of the computer revolution.

We could ask for no better equipment for tackling the challenges ahead of us than an attitude of informed and balanced skepticism about the impact of high technology on our right to be let alone. Thomas Jefferson once observed that "the ground of liberty must be gained by inches." Our liberty can be lost the same way—inch by inch. Perhaps the best way to ensure that our children do not awake one day to find that "Big Brother is watching you," is to encourage each of us to vigilance—each must carefully, continually, and confidently watch Big Brother back.

13

High Technology
and Higher Education

JAMES W. CAREY

Among the articles of the national faith, few have stronger resonance
than our belief in the beneficence of technology and the efficacy of
education. The beliefs were twin-born. Technology, freed from en-
crustation of the old world, purged of its bondage to old cities, old
elites, and older ways, set down in the garden of America, has always
promised us a general redemption: freedom from want, freedom
from weakness and corruption, freedom for a better life of peace,
prosperity, and plenitude. The twin pillar of education reinforced the
commitment to technology for it guaranteed not only the knowledge
with which to carry forward the technological project but underwrote, as well, the viability of a republican way of life; it ensured the
knowledge to make us free.

While technology and education have always underwritten the
quest for power, prosperity, and profit, they also underwrote the
dream of a democratic republic. They were institutions designed to
hold together, despite classical democratic theory, a large land and a
large population, and to render a people sufficiently educated that
they might govern themselves. This dream of an educational and
technological democracy has often been the equal of the dream for
wealth and power but of late the tables have turned. Technology and

An earlier version of this chapter appeared in *Illinois Issues*, March 1984. Copyright
Illinois Issues, published by Sangamon State University, Springfield, IL 62708.

education have been shorn of democratic pretension. The purpose of education, like that of technology is to make us richer and more powerful, more successful and more leisured, but hardly more equitable or more community centered, hardly more democratic or public spirited. That part of the national dream is back under the night of the republic.

When our schools are today singled out for their mediocrity, their failure to emphasize science, computer science, and technology, their failure to integrate themselves with the corporate and governmental sectors, you can be sure this rendition of our troubles has but one purpose: to regain a military edge against the Russians, a competitive edge against the Japanese, and a technological edge against everyone. The dream of an educated populace has largely faded from view and is often held in contempt. Beyond a few gestures at equality, gestures aimed at keeping the streets clear, education has no relevance to the political order, to public life. As Jonathan Kozol has recently demonstrated, education is no longer supposed to transmit even the minimum competence or literacy necessary for political participation. Corporations are the new partners of education, but it is far from a benevolent or disinterested patronage. The *quid pro quo* is a new labor force—docile but ambitious—and a new entrepreneurial spirit in academic research. With minor exceptions here and there, the federal government's interest in higher education has been reduced to military research.

This is the background against which to assess high technology, which, in state after state, has emerged as a savior, a mythic panacea for a variety of economic, educational, and political woes. The new glamour firms in electronics, computers, communications, robotics, and genetic engineering, firms which seem to be in infinite supply, promise everywhere to provide a cornucopia of jobs, markets, and products, to rejuvenate ailing economies, to refund declining universities, to reemploy the unemployed and redundant, to offer vast satisfying opportunities to those new to the labor force, to produce environmental harmony as high tech displaces the smokestacks of low tech, and even to eliminate, through user friendliness, the last alienation and estrangement between people and their machines.

Everywhere high tech is putting higher education back into prominence as the key institution where personnel for the new industries are to be trained and where research will yield the products and processes of the new world aborning. Indeed, it is sometimes hard to imagine that universities might have a purpose other than serving the needs of high-tech industry.

Despite our national penchant for founding a New Jerusalem or discovering a Passage to India, Americans take themselves to be a practical and hard-headed people. Yet, on the subject of high tech, a veritable rhetoric of the cybernetic sublime overtakes the calmest of minds and the most down to earth of our professionals, the engineers. Translated into advertising this rhetoric appeared over the logo of Apple Computers the week following election day, 1984:

> Last Tuesday, several million of you demonstrated the principle of democracy as it applies to politics. One person, one vote.
> Throughout this magazine, we're going to demonstrate the principle of democracy as it applies to technology. One person, one computer.

Despite the degraded view of democracy contained in the ad, one must admire the audacity of it all, the sheer hucksterism, the renewed booster spirit, the evangelical fervor and enthusiasm which promotes this latest generation of machinery. But this is hardly a plan; it is rather a talisman, a high-tech version of encyclopedia sales.

I am not trying to play the skeleton at the feast. One must give these proposals their due. The industrial activities represented by high tech are among the more promising opportunities for economic development now available. The United States possesses, and legitimately seeks to enhance and protect, certain competitive advantages in international trade in computers, communications, and other high-tech fields. There are opportunities for expanded employment in these industries, though for an all too narrow slice of the population.

But even if we grant this, it is still necessary to be skeptical about the high-technology proposals. These proposals have not erupted in every statehouse and major newspaper as a result of spontaneous, independent discoveries. In America the spontaneous is always planned. The campaign for "high-tech" is not merely an attempt to make the economy more efficient and competitive, albeit with a generous public subsidy. High technology is part of an attempt to change the direction of American life, to restructure American society. The offhand references to the needs of high-tech industry for a benign human environment, less restrictive social legislation and less militant labor unions signal more than the class bias of the proposals. These demands, and the frenzied competition they set off among the states, can be read as demands for pastoral places for the upper middle class to work free from the intrusion of the poor and disadvantaged, the absence of even minimal government regulation and the elimination of trade unions.

Moreover, the proposals for high technology are not exactly new. We are walking the furrows of ground plowed for the last two decades. These plans have been put forward by prognosticators and prophets who see in a new generation of machines another technological solution to what are in fact persistent political problems. Books on future shock, megatrends, the third wave, the computer society merely manifest in popular culture a vision of a desirable future loosely shared by a variety of groups: the major engineering societies, leading corporations with global stakes in high tech, universities looking for substitutes for declining federal support, the military seeking to augment its share of the gross national product, and the State Department searching for new technological means to maintain an American hegemony. Educated elites in turn pick up the theme that our competitive failing results from a widespread scientific illiteracy and propose, as with the Sloan Foundation, a new definition of the liberal arts emphasizing mathematics, computer science, and technological expertise. Anxious middle-class parents, eager to purchase a place for their children in the occupational structure, pack them off to computer camps or direct them even earlier toward Harvard via infant training at the personal home computer. The advertising of computer companies resurrects the oldest image of the literate man and weds him to the new computation devices: the priesthood of all believers, every man a priest with his own Bible, becomes in the new rendition the priesthood of all computers, every man a prophet with his own machine to keep him in control.

A new Morrill Act has been proposed to resuscitate the land-grant tradition. This one would forge a new partnership between business, the state, and the university, creating a national policy in which education is harnessed to real economic needs and learning is tied to a national strategy for economic growth. The new Morrill Act is a device whereby education can once again become a national priority and lay claim to increasing government resources because it is the key to revitalizing the private sector and the military.

Again, this is not, by indirection or innuendo, an argument against high technology. We probably need a good deal more of it than we are getting. Rather it is to question the proposals which isolate one sector of the economy for special public favor, which invest inordinate hopes and educational resources in one particular arena, and which concentrate attention on the needs of one narrow segment of the population. If such high-tech plans had been announced in the more politicized climate of the 1960s, they would have raised a specter. The partnership now bravely envisioned would have been quickly seen as a proposal on behalf of the military-industrial complex augmented by the collaboration of government and education: in potential, a vast new concentration of power. High technology would have been denounced as a code phrase for restoring the *status quo ante* after decades of liberal erosion of class perogatives. Most of all, it would have been seen as an attempt to restructure society along the lines demanded by the now homeless multinational corporation and to eradicate in the process the sometimes independence of education, the one institution that might occasionally march to a different drummer.

I begin this essay on a grim note—a note in which conspiracy is in the air and cabals of power are working in concert—not because I think high-tech proposals are without merit and certainly not because I think higher education is beyond reform.

I just don't think that "high tech," which seems to be the only wheel in town, the only rising star to which everyone is pinning their hopes, is adequate to the tasks we face either in the economy or in education. More importantly, I think it is a class-biased set of proposals of a particularly vicious kind which, unless augmented by other efforts at educational and economic development, will only harden and polarize already abrasive and destructive social divisions.

Under the circumstances, it seems necessary to ask, at the least, just what high technology is going to do for the country and what, more importantly, it is going to do for higher education. While this is a melancholy undertaking, we might review something of the history of higher education, particularly in its modern phase, if only as prelude to a conclusion.

Among the issues that dominated the troubled 1960s—a political era stretching from the first of the civil rights sit-ins to the Vietnam peace accords—were the attempts to make the universities more responsive to issues of equality and social justice, particularly for racial minorities, and to make the university less responsive to the needs of the national defense state, to the research and educational agenda of the federal government. It was a difficult line to walk. Those who attempted to walk it did so by emphasizing that originally universities had walls around them, walls that were both architectural and symbolic. They served to wall off the university from its surrounding community, to grant it special privilege and insulation from the undue influence of the larger community. Academic freedom, on this reading, was not an individual right of faculty members but a corporate right of the university to conduct its own affairs by its own lights, to pursue an intellectual agenda determined by its own scholarly priorities not by the short-run interests and needs of society. But as the walls held the society out, they also encircled the academic community and constituted a symbolic parentheses constraining academic life and granting it a special tone, rhythm, and pertinence. The Vietnam War revealed that the walls had been decisively breached, and the university seemed little more than the servant of outside interests. With that in mind many attempted to reassert the independence and integrity of the university tradition, of its right to pursue an order of values determined by its several subject matters, not by the demands of the professions, business, and the state, though it inevitably served those interests indirectly. At the same time it seemed perfectly consonant with the land-grant tradition to insist that space within the university ought to be available to all with the talents and interests to partake of it, that it ought not to be limited to those inscribed with the correct class background. There was a contradiction here, admittedly, and it is one that is unresolvable. It is precisely the ability to manage this contradiction—to defend the independence of the university tradition and still to make the institution responsive to fundamental values of the founding society—that is at the heart of university administration.

It takes, of course, a certain studied irrelevance to talk about walls encircling the U.S. public university since the Morrill Act of 1862, the founding legislation for most of them, had as its announced purpose the breaching of the walls of the traditional university. The legislation offered grants of land to each state for

> the endowment, support and maintenance of at least one college, where the leading object shall be . . . to teach such branches of learning as are related to agriculture and the mechanic arts, in such a manner as the legislatures of the states may respectively prescribe in order to promote the liberal and practical education of the industrial classes in the several pursuits and professions in life.

As the distinguished British educator Eric Ashby has commented, the dismantling of "the walls around the campus" was and is "the great American contribution to higher education. . . . When President Van Heis of Wisconsin said the boundaries of the university are the boundaries of the state, he was putting into words one of those rare innovations in the history of the university. It is one that has already been vindicated by history. Other nations are now beginning to follow the American example."

While I agree with Professor Ashby 's general assessment, it should also be emphasized that no sooner were the walls pulled down than they were put back up around the commonwealth of learning, around the liberal arts as the center of education. Those arts aimed not at practical mastery but at the cultivation of virtue, character, and the arts of living well. Such insulation not only protected the core of humane learning and prevented the university from being merged into the community, it also imposed a special and studied irrelevance to the core of university life. The founding legislation emphasized, as well, service of the university to individuals—farmers and mechanics—and to the industrial classes.

Late in the nineteenth century the walls of university life were driven down as the institution itself reached outward and the business and professional community reached inward. The university became infected, as Thorstein Veblen noted in *The Higher Learning in America*, with the spirit of business enterprise perhaps first and most continuously in the business of intercollegiate athletics. But business enterprise reached in, as Leland Stanford, John Rockefeller, Ezra Cornell, and other tycoons supplied the capital needed to support expansion and to hasten the transformation of the college into a university. This was done principally in private education, but the expansion and transformation set the example for public universities as well. As the connections grew, business leaders replaced clergy on the boards of trustees of such universities. And, naturally, major contributors thought they should have a say about what was taught and on the social attitudes and philosophy of the faculty and student body.

The professions and professional accrediting agencies also had a hand in the transformation of the universities. The professions desired the prestige of attachment with it, but they also sought to control the university and to bend it to professional purposes. The control over the curriculum demanded by accrediting agencies often ceded the intellectual tradition to professional societies and the interest groups behind them, while borrowing at the same time the mantle of difference and independence which had been the symbol of the university tradition.

These professional connections and the growing spirit of business enterprise came about long before the faculty became experts at seeking grants and independent funding. However, the presence of the federal government and a war economy from the late 1930s forward cemented the configuration of the new university. The success of academicians, if not the academy, in the development not only of weaponry but the entire social and behavioral technology of warfare gave birth in the postwar years to what Clark Kerr described as the multiversity, an institution branching out and radiating into every nook and cranny of the society, offering expertise, intelligence and training in every conceivable sphere of activity: a university without walls and without restraint. The formation of the multiversity in the years after World War II occurred as campuses were gradually opened to the mass of the population. This opening proceeded less from a democratic instinct than from the needs of an expanding economy and the worldwide spread of American industry. In 1947 a presidential commission reported that because of advancing technology there was a rapidly growing need for college-trained professional workers in the distributive and service occupations. Universities undertook this training as programs, departments and majors proliferated, and at the same time, research laboratories that are both technical and defense oriented in both the social and natural sciences were moved from the federal government to the universities. All this was overlaid on the traditional forms of liberal education, though that tradition became increasingly irrelevant, except in rhetoric, to the real purposes of the institution. Those subjects which defined the social and political purposes of education and therefore gave life to notions of virtue and character continued to be taught, of course, but they ceased to have much to do with what the institution was all about.

The growing contradictions of the multiversity, contradictions among its narrowly professional objectives, its mission-oriented research, and its foundations in the traditional disciplines, burst into the open during the 1960s. For a while it seemed the campuses were undergoing a revolution, though it was a "revolution with a taste in wine," as Norman Mailer described it at the time. Whether it was the wine or the music, the 1960s were heady, euphoric, and disorienting days, at least for those on the Left. Normalcy returned rather quickly in tribute to just how heady those days were, but one ought not to overlook the real gains that were made. The civil rights movement produced a perhaps irreversible change in the character of race relations and the status of black Americans, even if the economic gains of minorities have been overestimated. While the troubles on the campus did not exactly bring the war in Vietnam to an end, they did at least call attention to the deep involvement of the universities in the most questionable types of military and defense research, indeed to the alacrity with which the research establishment was willing to abandon all pretenses to a university tradition when it came to assuring sources of funding. There were other gains. The best of the students, at least for a while, queued up to the most "irrelevant" of the disciplines. Certain areas of the traditional liberal arts enjoyed a season in the sun, and professional schools in law and journalism were imbued with a sense of mission beyond the provision of the tools for individual success and social enhancement.

The events of the 1960s, however, did severely weaken U.S. higher education. While the system remained fiscally sound throughout the 1960s, even as new four-year and senior institutions were created and the community college system brought into existence, its financial base decayed throughout the 1970s and the early years of this decade. Research support was gradually withdrawn by the federal government. It left because of the proliferation of independent research firms that grew up to siphon off the federal largesse and because the university proved to be less efficient, less reliable, and more porous to disruption than research firms in the private sector. But state legislatures, now grown used to the federal government underwriting education, also failed to provide support commensurate with inflation, so that new building and maintenance ground to a halt, and a faculty, which had abandoned vows of poverty during the good years, anxiously watched its status and its income eaten away. The decline in public support had something to do, of course, with the rupture in loyalties that were provoked by the Vietnam demonstrations and some bizarre students that accompanied them. But it also had something to do with the belief that higher education had joined the political process as one more interest group, whose interests in the public weal extended about as far as its own economic security. This change was as much demanded by universities as forced upon them by legislatures. All the apparatus of bureaucracy once reserved for roads and sewers were applied to education: formula funding, student head counts, square footage calculation. For a while it was advantageous, particularly given a rising educational market. In the end a destructive bargain was struck: the university became judged on one side by its contribution to the economy—trained personnel and useful knowledge—and on the other by its conformity to the arithmetic of support. The calculus left no room any longer for thinking of the university as different from banks or corporations or athletic departments or economic development agencies. When notions of an educated citizenry, a democratic populace, and a cultivated way of life evaporated as objects of education, the university was seen as merely another participant in the process of interest group politics and not an institution representing a public and indivisible interest with the general welfare.

The point of this review is that the problems of higher education, as of the culture generally, go much deeper than and are not susceptible to cure by high technology. Indeed, what high technology principally offers is a new political coalition to shore up the otherwise declining support of higher education. But the terms on which this support is sought would further reduce the independence of the university, further confuse the values of the student, and further distort the aims of education rightly conceived: the creation of an enlightened and cultivated citizen. The deeper confusions in higher education simply are not touched by the solutions of high technology. But does high technology offer solutions to the wider array of social problems or to the rejuvenation of the U.S. economy?

Here, too, I think we can be skeptical. Skepticism was more apparent when the romance of high technology first appeared during the Vietnam War period. A series of prophetic voices in the mid 1960s proclaimed a technological revolution to be realized through the marriage of computers and television, communications and information processing: a new stage of economic development, the postindustrial society, the information society, the postmodern society. Also, there were the various commissions and study groups that prepared us for the future, for the twenty-first century. The development of this form of talk was not without precedent. At a similar moment in the nineteenth century a new generation of machines was heralded as the device through which the past would be abolished and a new world created. In the wake of the Civil War, in the long depression that dominated the period 1873–93, the new devices of electricity seemed to be the means by which not only a new economy but indeed a new social order would be fashioned. The telephone and the electric generator promised the capacity to wed together the myth and the powerhouse, the dynamo and the virgin, the promise of peace, harmony, and culture with the drive to power, profit, and productivity. This rhetoric of an electrical sublime was resuscitated and given new vigor in the 1960s. It was now a cybernetic sublime, a new era of electronics, communications, and computers. The new generation of machines promised a new direction in American life—beyond the devastations of the past and present into a brighter lit world of a global village on spaceship earth: an era of economic power, ecological balance, and social justice. The point is simply this: The current romance with high technology, insofar as it exceeds rational expectation, is but an extension of prophecy that resurfaced in the 1960s. It is part of the American religion of technology.

But there was a parallel development in the 1960s, namely a shift in the intellectual underpinnings of American thought from Left to Right. Recent developments in education and politics have not come about automatically or inexorably but are the result of steady intellectual work, work that responded not only to liberal policies of the 1960s and the Vietnam war, but to the entire climate of ideas that dominated that political epoch. I am speaking, of course, of the rise of neoconservatism. Its emergence can be precisely dated to the founding of the journal *The Public Interest* in 1965. While it is easy to identify neoconservatism with specific issues—affirmative action, the resurgence of the Cold War and anti-Soviet sentiments, the reaction to the women's movement—it is important to see it in a wider context. Neoconservatism has pasted together issues and positions across a broad front. The pages of *The Public Interest* and other neoconservative journals contain essays on a wide variety of issues: the environment, crime, welfare, education, technology, the political parties, population, natural resources.

Neoconservatism has become the decisive political ideology. Not only has it provided the intellectual underpinnings of Reaganism, but it has allowed conservatism to escape being merely the soured wine of religious fundamentalism and to articulate a sense of a future that could embrace the deep contradictions and antinomies of American life. The attitudes toward high technology and education within neoconservatism are more than a mere nativism; they are part of a vision which attempts to capture and preserve what is useful and complex in American culture, to consolidate the gains of the recent past and to project those gains into a future as an adaptation to the conditions of modern politics and the economy. Its achievement should not be underrated even while its weaknesses are noted. America is still badly divided by interest groups. The fissures on both sides of the political spectrum are sharp and deep, and there is hardly a consensus on anything. Nonetheless, neoconservatism has allowed a center to form, however weak and temporary, that contains these fissures and produces a broad social program that provides an ideological base and direction to U.S. politics and institutions.

Beyond its response to 1960s movements surrounding Vietnam and civil rights, neoconservatism has attempted to deal with the vexing economic problems which the war revealed or decisively exacerbated: an inflation that resulted in significant part from fighting a war without declaring it and financing a war without legislating it; an economy which was insufficiently militarized for the tasks assigned it, so tuned to current consumption that it could not save enough to finance its continued growth and modernization, and so dependent upon oil that only a continued U.S. international hegemony could stabilize energy cost; and a level of inefficiency in basic industries that reached scandalous proportions.

The marriage, then, of higher technology and higher education is not to be analyzed or judged merely as a temporary expedient to solve this problem or that. It is not merely a way of dealing with inflation or ending unemployment or revitalizing the economy. Neither is it merely a temporary justification for refunding the American university or securing a new basis of research support or providing adequate facilities so that students can major in engineering or business or whatever else they may wish. Beyond these obvious and reasonable goals, the "high tech–higher ed" merger is part of the larger neoconservative proposal for a new way of life, one which repudiates the recent past in order to envision a new phase of American capitalism. As presented by and to universities, the high-tech solution masks its more conservative assumptions. It is presented, advertised, and sold as a total solution when it is at the very best merely a partial one. Moreover, it is a technical solution that disguises a political one. It proposes to solve basic social contradictions via the development of technology, but it does not contain solutions to the fundamental problems of either higher education or the economy.

It is assumed, for example, that the growth of high-technology industry will lead to a rapid expansion of professional and technical occupations that require considerable education in computer-related areas and that high technology will require upgraded skills because workers will be using computers and other technical equipment. However, as a recent report in MIT's publication, *Technology Review*, put it, "these assumptions are dead wrong." There are not going to be enough "high-tech" jobs to replace jobs lost because of declining industries. While unemployment for engineers and computer specialists will grow almost three times as fast as employment overall, these occupations will generate only about 7 percent of all new jobs during the rest of this decade. Employment for computer systems analysts will increase by over 100 percent between 1978 and 1990 and still add only 200,000 new jobs to the work force. The great expansion in jobs in the years ahead will be in low-skill and low-paid areas, while many other jobs will be downgraded in the skills required because of automation. Further, vast numbers of those in the middle levels of employment will have their jobs absorbed into the computing capacity of the new machines. In short, the growth of high technology is not an answer to American employment problems nor an enhancement of the conditions of work of most. Skilled jobs will continue to be scarce and will continue to be exported to countries of the Pacific Rim. The tendency toward a two-tier level of employment will increase: a relatively few high-paid executives, engineers, and scientists and an increasingly proletarian work force. This merely recapitulates the recent history of the effect of technology on employment and, as the *Technology Review* piece concludes, should "revise educational priorities and place greater emphasis on a strong general education rather than a narrow specialized one."

The high-tech solution is simply irrelevant to the vast majority of American workers, particularly the currently unemployed. The campaign for a narrowly based high-tech policy is a campaign that benefits one particular class. There is perhaps nothing wrong with that,

194

but what is questionable is the assumption that the cost of scientific and technological development, costs which ought properly to be borne by the corporate community, should be put on the public budgets of individual states when the resources produced—knowledge, talent, ideas, individuals—are mobile and free to migrate wherever conditions are more benign in or out of the country.

The second questionable assumption of high-tech policy, and one that merely complements the employment assumption, is that traditional U.S. industries of steel and automobiles, for example, are dead, that they were killed by high labor costs and that there is no reviving them. Given this verdict, there seem to be no viable options other than an evacuation of traditional manufacturing and a complete move into the newer, more glamorous high-tech industries based on robotics or other labor saving devices, which have lower labor costs. Although wages may be lower for laborers, high-tech firms offer employment to scientists, engineers, and the white-collar class, generally at good salaries. Again, this scenario virtually ignores the majority of the working population. It also conveniently places blame for the failure of American industry on the American worker, thereby justifying a policy of reducing wages and wage labor while increasing administrative costs and salaries. But basic U.S. industries have been afflicted by chronically bad management, and high-tech firms are not immune to the same disease. The higher cost of much U.S. industry derives not only from frequent overpayment to wage labor, but from bloated administrative salaries, bloated in both absolute terms and as a multiple of what is paid ordinary workers. Even worse, bloated administrative salaries and other perquisites have insulated management from the actual demands of the market and the actual performance of products. When such markets were protected, little trouble occurred, but at the first hint of competition such firms were unable to adjust their administrative cost structure or management practices.

But the blindspot to the needs of reindustrialization derives from a deeper assumption about the "postindustrial" or "information society," namely that we are to become a software economy rather than an industrial one. In some versions we are to process French banking data, write computer routines, make programs for satellite broadcasting, and generally take in one another's washing. However, the manufacture of the washing machines, computers, and other hardware of the new economy will be left to other countries, principally to Japan and other nations of the Pacific Rim. A software economy, one that exports all the basic trades and crafts, will be, I believe, a cultural disaster. More to the point, such an economy assumes the permanent existence of a large, growing, surplus population living on enforced leisure or the work-welfare leftovers of an upper-middle-class work force.

There are, then, intrinsic limitations to high-technology solutions to our economic and educational problems. Moreover the problems of both education and the economy are broadly cultural rather than narrowly economic. They derive from what the neoconservative writer Daniel Bell has called the cultural contradictions of capitalism. Higher education would have a more salutary role in dealing with these contradictions if it put forward general strategies of reindustrialization that went beyond projections of narrow and class-based interests. No high-tech policy will be effective that does not lead to an overhaul of management practices and that does not include a general plan for reindustrialization rather than merely favor growth of a few selected glamour industries. While the word reindustrialization evokes the image of Felix Rohatyn's plans for bailing out failing and uncompetitive firms, it need not include that. Indeed, it has become increasingly clear that public investment in industry in both Europe and Japan is the margin by which many of our basic industries are losing ground. I am not here arguing for such public investment, though I think it deserves more consideration, but for some strategy of reindustrialization that does more than abandon basic industries, basic workers and their basic skills, particularly if it is to be underwritten by public monies channeled through universities or elsewhere.

Finally, both the society and education would benefit from a less utilitarian and longer term approach to higher education. In the midst of the Depression, Walter Lippmann gave some unsolicited advice about the "role of scholars in a troubled world." In the midst of such troubles the scholar feels he should be doing something about them or at least be saying something that will help others do something about them. "The world needs ideas; how can he sit silently in his study and with a good conscience go on with his thinking when there is so much that urgently needs to be done?" But it is precisely this strategy of staying in the study that he recommends. This view, he concludes, will seem to many a mere elegy to a fugitive and cloistered virtue:

> Yet I doubt whether the student can do a greater work for his nation in this grave moment of its history than to detach himself from its preoccupations, refusing to let himself be absorbed by distractions about which, as a scholar, he can do almost nothing. For this is not the last crisis in human affairs. The world will go on somehow and more crises will follow. It will go on best, however, if among us there are men who have stood apart, refused to be anxious or too much concerned, who were cool and inquir-

ing, had their eyes on a longer past and a longer future. (Lippmann 515)

Lippmann's defense of the university tradition in the midst of a grave economic crisis is an essentially conservative one. It was later matched by a defender of capitalism, conservative economist Joseph Schumpeter. In his defense, which I here twist, Schumpeter ruefully acknowledged the tendency of capitalism to absorb and destroy—to "creatively destroy"—the institutional framework which guarantees its continuance: in mythological terms, to eat its children as it banishes its parents. Capitalism, in his view, could only be successful within a protecting framework made from noncapitalist material. Without protection by noncapitalist institutions, many of which, like the university, were essentially medieval in origin, capitalism is "politically helpless" and even unable to care for its own interests. The tendency of capitalist institutions to live solely at the horizon of this quarter's profits or next year's opportunity costs and to absorb all other institutions into the principles of the market eats away both psychologically and institutionally the resources upon which the system is raised. Therefore, the encapsulating framework of noncapitalist habits of thought and life are indispensable for the tenacity, creativity, and prudence necessary to keep a capitalist society functioning effectively (Schumpeter).

William Morris once said that the most important product of the mine was the miner. Similarly, the most important product of education is not knowledge and certainly not a knowledge industry; nor is it research, or service, or a work force. The most important product of education is the student. Everything a university does cultivates a student, and therefore universities teach by their presence, by the values they represent in every act.

Lippmann and Schumpeter suggested that the one indispensable lesson of universities is the ability to stand apart, to be cool and inquiring, to detach from the immediate and short run demands of the market in order to take a longer view of the past and future. This is the core of university teaching and the indispensable gift such institutions have to contribute to society. Because today's students will be tomorrow's elites, the key question comes down to the kinds of elites we are creating in the fevered atmosphere of high technology, quick fixes, and short-run solutions. One reason that American life has been unsatisfactory in recent years is that our elites have been unable to rise above the immediate values of the market, to have any sense of collective identification and obligation, to be more than mere professionals. The criterion for membership in that elite is short-run success in the climb, the morally neutral ability to perform. The new plans for high tech education threaten to convert the entire society into a meritocracy and technocracy and to extirpate any residual sense of a common and collective life that extends beyond the moment or the morrow. The core of humane learning is the possession of a common culture with enough durability to transform today's students into tomorrow's leaders, persons who feel a sense of care and responsibility for their fellow citizens and for the noblest of our democratic traditions. That is still the essence of university life and any humane program of education even in the age of high-tech.

Works Cited

Kozol, Jonathon
  1985   *Illiterate America.* Garden City, N.Y.: Doubleday.
Lippmann, Walter
  1963   *The Essential Lippmann.* Edited by Clinton Rossiter and James Lare. New York: Random House.
Schumpeter, Joseph
  1942   *Capitalism, Socialism and Democracy.* New York: Harper and Brothers.
Veblen, Thorstein
  1918   *The Higher Learning in America.* New York: B.W. Handbook.

INDEX

Clarke, J. M. F., 118, 120, 130
Coleridge, S., 31
Collective bargaining, 14–15, 106–11; by communities, 145–55
Community: of celebration, 171–74; defined, 8, 117, 133; impact of technology on, 135, 139–40, 142; and planning, 145, 149–50, 154; redirecting technology through, 117–29, 151–55; "repaired" vs. "redemption" of, 12, 161–62, 167–68; and subsidiary communities, 16
Computer matching, 179–80
Computer technology, 176–82; as cognitive model, 60–62; and mind, 89; and personal computers, 84; surveillance by, 8–9. *See also* Technological surveillance
Condorcet, M., 25, 87
Consciousness, 1, 8
Control: ideology of, 87; over production, 83, 92–93, 137–38; over reality, 84
Cooke, R., 7, 16
Cooley, D., 118–19, 131
Copernicus, N., 61
Corbett, T., 122, 130
Cornell, E., 189
Cunningham, R., 123, 130
Cyert, R., 105

Dahl, R., 101–2, 169
Darlins, R. C., 129
Davis, F., 41, 51
Davison, J. R., 129
Decentralization, 171, 174
Deindustrialization, 15, 105
DeLillo, D., 11–12, 17
Democracy, 12; and democratic technology, 84–87; and democratic values, 6, 12–14, 17; and yeoman democracy, 14
Descartes, R., 3, 31, 87
Deskilling, 84, 93
DeVries, W., 125, 128

Disney, W., 10–11, 46–49
Disneyland, 46–47, 50
Disneyworld, 46–47, 50
Dixon, J., 118–20, 130
Donoghue, D., 44, 51
Dornette, W., 118, 130
Douglas, A., 5, 17, 45, 51
Drucker, P., 81, 85, 90, 98, 102

Eckhous, A., 129
Economic development, 151–55
Economics, 92; and dumbbell economies, 15; Keynesian, 85
Edison, T., 9, 11, 30, 44–45
Ehmer, J., 141, 143
Eisenhower, D., 105
Ellul, J., 4, 18
Emerson, R., 3, 5, 10, 30, 32
Enlightenment, 2–3, 6, 25–26, 28, 34
Epcot Center, 46
Everett, E., 28

Factory system, 26
Fallows, J., 15, 18
Farr, A. D., 118–19, 121, 130
Farrell, J., 45
Ferguson, T., 164, 174
Feudalism, 9, 25, 40
Fichte, J., 31
Finch, C., 48–49, 51
Findley, L. J., 118, 130
Firnberg, H., 133, 141, 143
Fischer, W., 122–23, 131
Fisher, M., 24, 36
Flexible specialization, 8, 14, 86–87
Flexible systems of production, 95–96, 135
Ford, G., 105
Ford, H., 10–11, 38, 43–48, 50; and "Fordism," 31
Foster, S., 45
Frank, S., 126
Frankfurt School, 26
Franklin, B., 25, 27–29, 35, 36, 38
Fredkin, E., 89

Freeman, R., 100–102
Fulton, R., 38
Furman, E. B., 131

Galileo, 2–3, 31, 62–63
Geertz, C., 8, 18
General Motors, 15, 96, 108–10
Goldberger, P., 47–48, 51
Gonzalez, E., 119–20, 130
Goodman, R., 155
Gore, A., 125
Gorner, P., 126–27, 130
Granbois, J., 121, 131
Greenfield Village, 44, 46, 50
Gutman, H., 13, 18, 136, 143

Haas, C., 46–47, 51
Habermas, J., 26, 36
Hai, D., 122, 130
Halper, T., 121–23, 130
Hamilton, A., 13
Hareven, T., 137, 143
Harris, L., 182
Harris, N., 42
Harris, T. J., 119, 130
Harvey, W., 31
Haugeland, J., 55, 62–63
Hawthorne, N., 5, 32–33
Haydon, M., 124
Hench, J., 46–47
Henry V, 177
Herteaux, M., 134, 143
High technology, 145–46, 148, 150–
54, 178, 182, 184–87, 191–97
History: as restoration, 44–45;
as succession of styles, 45; as
technological progress, 29, 37;
as tradition, 38
Horkheimer, M., 26, 36
Hospice movement, 16, 117, 120–24
Hounder, S., 123, 130
Hounshell, D., 39, 51
Humanities, 4–7, 12–13; and public
discourse, 4–7; and relativism, 5;
traditions of, 3, 6. *See also* Lib-
eral arts

Iglehart, J., 126, 130
Industrial Revolution, 5, 25, 31, 136–
38, 140–41. *See also* Second Industrial Revolution
Irving, W., 41

Jacobs, J., 7, 18
James, H., 38, 41, 51
Jardim, A., 45, 51
Jarvik, R., 125–26
Jazz Age, 43
Jefferson, T., 10, 13, 25–29, 35–36,
166, 182
Jehovah's Witnesses, 7, 16, 117–21,
124
Jeremiad, 39
Johnson, L. B., 49, 105
Jones, R., 126–27, 130

Kalin, M., 3, 11
Kant, I., 31
Kaplan, J., 41, 51
Kasson, J., 5, 8–10, 13, 18, 41, 52,
166, 174
Kemmis, D., 159
Kennedy, J., 49, 105
Kepler, J., 31
Kerr, C., 189
Keynes, J. M., 85, 97
Kozol, J., 184, 198
Krant, M., 122, 131
Kuttner, R., 86, 90

Lack, S., 122–23, 131
Lapic, R., 119
Lasch, C., 4–5, 8, 13–14, 18, 26
Leaver, E. W., 82, 90
Lelyveld, J., 49, 52
Leonteif, W., 99–100, 102
Liberal Arts, 13; and cultivation of
virtue, 188–89, 191, 198
Liberalism, 7, 159–61, 164, 167–68,
172–73; crisis of, 159, 161–63
Lincoln, A., 48–49
Linowes, D., 178